Mining Heterogeneous
Information Networks

Principles and Methodologies

Synthesis Lectures on Data Mining and Knowledge Discovery

Editors

Jiawei Han, *University of Illinois at Urbana-Champaign*
Lise Getoor, *University of Maryland*
Wei Wang, *University of North Carolina, Chapel Hill*
Johannes Gehrke, *Cornell University*
Robert Grossman, *University of Chicago*

Synthesis Lectures on Data Mining and Knowledge Discovery is edited by Jiawei Han, Lise Getoor, Wei Wang, Johannes Gehrke, and Robert Grossman. The series publishes 50- to 150-page publications on topics pertaining to data mining, web mining, text mining, and knowledge discovery, including tutorials and case studies. The scope will largely follow the purview of premier computer science conferences, such as KDD. Potential topics include, but not limited to, data mining algorithms, innovative data mining applications, data mining systems, mining text, web and semi-structured data, high performance and parallel/distributed data mining, data mining standards, data mining and knowledge discovery framework and process, data mining foundations, mining data streams and sensor data, mining multi-media data, mining social networks and graph data, mining spatial and temporal data, pre-processing and post-processing in data mining, robust and scalable statistical methods, security, privacy, and adversarial data mining, visual data mining, visual analytics, and data visualization.

Mining Heterogeneous Information Networks: Principles and Methodologies
Yizhou Sun and Jiawei Han
2012

Mining Heterogeneous Information Networks: Principles and Methodologies

Yizhou Sun and Jiawei Han

ISBN: 978-3-031-00774-3   paperback
ISBN: 978-3-031-01902-9   ebook

DOI 10.1007/978-3-031-01902-9

A Publication in the Springer series
SYNTHESIS LECTURES ON DATA MINING AND KNOWLEDGE DISCOVERY

Lecture #5
Series Editors: Jiawei Han, *University of Illinois at Urbana-Champaign*
                Lise Getoor, *University of Maryland*
                Wei Wang, *University of North Carolina, Chapel Hill*
                Johannes Gehrke, *Cornell University*
                Robert Grossman, *University of Chicago*
Series ISSN
Synthesis Lectures on Data Mining and Knowledge Discovery
Print 2151-0067   Electronic 2151-0075

Mining Heterogeneous Information Networks

Principles and Methodologies

Yizhou Sun and Jiawei Han
University of Illinois at Urbana-Champaign

SYNTHESIS LECTURES ON DATA MINING AND KNOWLEDGE DISCOVERY #5

ABSTRACT

Real-world physical and abstract data objects are interconnected, forming gigantic, interconnected networks. By structuring these data objects and interactions between these objects into multiple types, such networks become *semi-structured heterogeneous information networks*. Most real-world applications that handle big data, including interconnected social media and social networks, scientific, engineering, or medical information systems, online e-commerce systems, and most database systems, can be structured into heterogeneous information networks. Therefore, effective analysis of large-scale heterogeneous information networks poses an interesting but critical challenge.

In this book, we investigate the principles and methodologies of mining heterogeneous information networks. Departing from many existing network models that view interconnected data as homogeneous graphs or networks, our semi-structured heterogeneous information network model leverages the rich semantics of typed nodes and links in a network and uncovers surprisingly rich knowledge from the network. This semi-structured heterogeneous network modeling leads to a series of new principles and powerful methodologies for mining interconnected data, including: (1) rank-based clustering and classification; (2) meta-path-based similarity search and mining; (3) relation strength-aware mining, and many other potential developments. This book introduces this new research frontier and points out some promising research directions.

KEYWORDS

information network mining, heterogeneous information networks, link analysis, clustering, classification, ranking, similarity search, relationship prediction, user-guided clustering, probabilistic models, real-world applications, efficient and scalable algorithms

Contents

Ming Ji

Department of Computer Science, University of Illinois at Urbana-Champaign

PART III Relation Strength-Aware Mining

6 Relation Strength-Aware Clustering with Incomplete Attributes

7 User-Guided Clustering via Meta-Path Selection

Acknowledgments

This research was supported in part by the U.S. Army Research Laboratory under Cooperative Agreement No. W911NF-09-2-0053 (NS-CTA) and W911NF-11-2-0086; MIAS, a DHS-IDS Center for Multimodal Information Access and Synthesis at UIUC; U.S. National Science Foundation grants IIS-0905215, CNS-0931975, CCF-0905014, and IIS-1017362; and U.S. Air Force Office of Scientific Research MURI award FA9550-08-1-0265. The views and conclusions contained in our research publications are those of the authors and should not be interpreted as representing any funding agencies.

We gratefully acknowledge Ming Ji for her contribution of *Chapter 3: Classification of Heterogeneous Information Networks* to this book.

Yizhou Sun and Jiawei Han
July 2012

CHAPTER 1

Introduction

We are living in an interconnected world. Most of data or informational objects, individual agents, groups, or components are interconnected or interact with each other, forming numerous, large, interconnected, and sophisticated networks. Without loss of generality, such interconnected networks are called *information networks* in this book. Examples of information networks include social networks, the World Wide Web, research publication networks [22], biological networks [55], highway networks [32], public health systems, electrical power grids, and so on. Clearly, information networks are ubiquitous and form a critical component of modern information infrastructure. The analysis of information networks, or their special kinds, such as social networks and the Web, has gained extremely wide attentions nowadays from researchers in computer science, social science, physics, economics, biology, and so on, with exciting discoveries and successful applications across all the disciplines.

In most of the current research on network science, social and information networks are usually assumed to be *homogeneous*, where nodes are objects of the same entity type (e.g., person) and links are relationships from the same relation type (e.g., friendship). Interesting results have been generated from such studies with numerous influential applications, such as the well-known PageRank algorithm [10] and community detection methods. However, most real-world networks are *heterogeneous*, where nodes and relations are of different types. For example, in a healthcare network, nodes can be patients, doctors, medical tests, diseases, medicines, hospitals, treatments, and so on. Treating all the nodes as of the same type may miss important semantic information. On the other hand, treating every node as of a distinct type may also lose valuable information. It is important to know that patients are of the same kind, comparing with some other kinds, such as doctors or diseases. Thus, *a typed, semi-structured heterogeneous network modeling may capture essential semantics of the real world.*

Typed, semi-structured heterogeneous information networks are ubiquitous. For example, the network of Facebook consists of persons as well as objects of other types, such as photos, posts, companies, movies, and so on; in addition to friendship between persons, there are relationships of other types, such as person-photo tagging relationships, person-movie liking relationships, person-post publishing relationships, post-post replying relationships, and so on. A university network may consist of several types of objects like students, professors, courses, and departments, as well as their interactions, such as teaching, course registration or departmental association relationships between objects. Similar kinds of examples are everywhere, from social media to scientific, engineering or medical systems, and to online e-commerce systems. Therefore, heterogeneous information networks are powerful and expressive representations of general real-world interactions between different kinds of network entities in diverse domains.

In this book, we investigate the principles and methodologies for mining heterogeneous information networks, by leveraging the semantic meaning of the types of nodes and links in a network, and propose models and algorithms that can exploit such rich semantics and solve real-world problems. Heterogeneous information networks often imply rather different semantic structures from that in homogeneous networks. Links in heterogeneous networks indicate the interactions between various types of objects in a network, and usually imply *similarity* or *influence* among these objects, that can be difficult to be expressed by traditional features. Information is propagated across various kinds of objects in a network, via various kinds of relationships (i.e., heterogeneous links), carrying different semantics and having different strengths in determining the "influence" across linked objects. These principles have laid the foundation for methodologies of handling various mining tasks in heterogeneous information networks, including ranking, clustering, classification, similarity search, relationship prediction, and relation strength learning. We will introduce these mining tasks and their associated new principles and methodologies chapter by chapter.

1.1 WHAT ARE HETEROGENEOUS INFORMATION NETWORKS?

An information network represents an abstraction of the real world, focusing on the *objects* and the *interactions* between the objects. It turns out that this level of abstraction has great power in not only representing and storing the essential information about the real world, but also providing a useful tool to mining knowledge from it, by exploring the power of links. Formally, we define an information network as follows.

Definition 1.1 (Information network) An *information network* is defined as a directed graph $G = (\mathcal{V}, \mathcal{E})$ with an object type mapping function $\tau : \mathcal{V} \rightarrow \mathcal{A}$ and a link type mapping function $\phi : \mathcal{E} \rightarrow \mathcal{R}$, where each object $v \in \mathcal{V}$ belongs to one particular object type $\tau(v) \in \mathcal{A}$, each link $e \in \mathcal{E}$ belongs to a particular relation $\phi(e) \in \mathcal{R}$, and if two links belong to the same relation type, the two links share the same starting object type as well as the ending object type.

Different from the traditional network definition, we explicitly distinguish object types and relationship types in the network. Note that, if a relation exists from type A to type B, denoted as $A\,R\,B$, the inverse relation R^{-1} holds naturally for $B\,R^{-1}\,A$. R and its inverse R^{-1} are usually not equal, unless the two types are the same and R is symmetric. When the types of objects $|\mathcal{A}| > 1$ or the types of relations $|\mathcal{R}| > 1$, the network is called **heterogeneous information network**; otherwise, it is a **homogeneous information network**.

Given a complex heterogeneous information network, it is necessary to provide its meta level (i.e., schema-level) description for better understanding the object types and link types in the network. Therefore, we propose the concept of network schema to describe the meta structure of a network.

Definition 1.2 (Network schema) The *network schema*, denoted as $T_G = (\mathcal{A}, \mathcal{R})$, is a meta template for a heterogeneous network $G = (\mathcal{V}, \mathcal{E})$ with the object type mapping $\tau : \mathcal{V} \to \mathcal{A}$ and the link mapping $\phi : \mathcal{E} \to \mathcal{R}$, which is a directed graph defined over object types $\mathcal{A}$, with edges as relations from $\mathcal{R}$.

The network schema of a heterogeneous information network has specified type constraints on the sets of objects and relationships between the objects. These constraints make a heterogeneous information network semi-structured, guiding the exploration of the semantics of the network.

Heterogeneous information networks can be constructed from many interconnected, large-scale datasets, ranging from social, scientific, engineering, to business applications. Here are a few examples of such networks.

1. **Bibliographic information network**. A bibliographic information network, such as the computer science bibliographic information network derived from DBLP, is a typical heterogeneous network, containing objects in four types of entities: *paper* (P), *venue* (i.e., conference/journal) (V), *author* (A), and *term* (T). For each paper $p \in P$, it has links to a set of authors, a venue, and a set of terms, belonging to a set of link types. It may also contain citation information for some papers, that is, these papers have links to a set of papers cited by the paper and a set of papers citing the paper.

   The network schema for a bibliographic network and an instance of such a network are shown in Figure 1.1.

2. **Twitter information network**. Twitter as a social media can also be considered as an information network, containing objects types such as *user, tweet, hashtag*, and *term*, and relation (or link) types such as *follow* between users, *post* between users and tweets, *reply* between tweets, *use* between tweets and terms, and *contain* between tweets and hashtags.

3. **Flickr information network**. The photo sharing website Flickr can be viewed as an information network, containing a set of object types: *image, user, tag, group*, and *comment*, and a set of relation types, such as *upload* between users and images, *contain* between images and tags, *belong to* between images and groups, *post* between users and comments, and *comment* between comments and images.

4. **Healthcare information network**. A healthcare system can be modeled as a healthcare information network, containing a set of object types, such as *doctor, patient, disease, treatment*, and *device*, and a set of relation types, such as *used-for* between treatments and diseases, *have* between patients and diseases, and *visit* between patients and doctors.

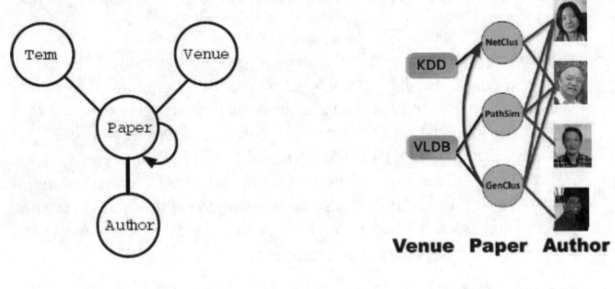

(a) Schema of a bibliographic network

(b) A bibliographic network

Figure 1.1: A bibliographic network schema and a bibliographic network instance following the schema (only papers, venues, and authors are shown).

Heterogeneous information networks can be constructed almost in any domain, such as social networks (e.g., Facebook), e-commerce (e.g., Amazon and eBay), online movie databases (e.g., IMDB), and numerous database applications. Heterogeneous information networks can also be constructed from text data, such as news collections, by entity and relationship extraction using natural language processing and other advanced techniques.

Diverse information can be associated with information networks. Attributes can be attached to the nodes or links in an information network. For example, location attributes, either categorical or numerical, are often associated with some users and tweets in a Twitter information network. Also, temporal information is often associated with nodes and links to reflect the dynamics of an information network. For example, in a bibliographic information network, new papers and authors emerge every year, as well as their associated links. Such issues will be addressed in some tasks on mining information networks.

1.2 WHY IS MINING HETEROGENEOUS NETWORKS A NEW GAME?

Numerous methods have been developed for the analysis of homogeneous information networks, especially on social networks [1], such as ranking, community detection, link prediction, and influence analysis. However, most of these methods cannot be directly applied to mining heterogeneous information networks. This is not only because heterogeneous links across entities of different types may carry rather different semantic meaning but also because a heterogeneous information network in general captures much richer information than its homogeneous network counterpart. A homogeneous information network is usually obtained by projection from a heterogeneous information network, but with significant information loss. For example, a co-author network can be obtained by projection on co-author information from a more complete heterogeneous bibliographic network. However, such projection will lose valuable information on what subjects and which papers the authors were collaborating on. Moreover, with rich heterogeneous information preserved in an original heterogeneous information network, many powerful and novel data mining functions can be developed to explore the rich information hidden in the heterogeneous links across entities.

Why is mining heterogeneous networks a new game? Clearly, information propagation across heterogeneous node and links can be very different from that across homogeneous nodes and links. Based on our research into mining heterogeneous information networks, especially our studies on ranking-based clustering [66; 69], ranking-based classification [30; 31], meta-path-based similarity search [65], relationship prediction [62; 63], relation strength learning [61; 67], and network evolution [68], we believe there are a set of new principles that may guide systematic analysis of heterogeneous information networks. We summarize these principles as follows.

1. **Information propagation across heterogeneous links.** Similar to most of the network analytic studies, links should be used for information propagation. However, the new game is *how to propagate information across heterogeneous types of nodes and links*, in particular, how to compute ranking scores, similarity scores, and clusters, and how to make good use of class labels, across heterogeneous nodes and links. No matter how we work out new, delicate measures, definitions, and methodologies, a golden principle is that *objects in the networks are interdependent and knowledge can only be mined using the holistic information in a network.*

2. **Search and mining by exploring network meta structures.** Different from homogeneous information networks where objects and links are being treated either as of the same type or as of un-typed nodes or links, heterogeneous information networks in our model are semi-structured and typed, that is, associated with nodes and links structured by a set of types, forming a network schema. The network schema provides a meta structure of the information network. It provides guidance of search and mining of the network and helps analyze and understand the semantic meaning of the objects and relations in the network. Meta-path-based similarity search and mining introduced in this book demonstrate the usefulness and the power of exploring network meta structures.

3. **User-guided exploration of information networks.** In a heterogeneous information network, there often exist numerous semantic relationships across multiple types of objects, carrying subtly different semantic meanings. A certain weighted combination of

relations or meta-paths may best fit a specific application for a particular user. Therefore, it is often desirable to automatically select the right relation (or meta-path) combinations with appropriate weights for a particular search or mining task based on user's guidance or feedback. User-guided or feedback-based network exploration is a useful strategy.

1.3 ORGANIZATION OF THE BOOK

The first chapter introduces the problem of mining heterogeneous information networks. After that, the book is organized into three parts, each containing two chapters that present principles and methodologies for mining heterogeneous information networks, organized according to different mining tasks. Finally, Chapter 8 outlines a few open research themes in this domain. The major contents of Chapters 2–7 are summarized as follows.

In *Part I: Ranking-Based Clustering and Classification*, we introduce several studies on basic mining tasks such as clustering and classification in heterogeneous information networks, by distinguishing the information propagation across different types of links.

- **Chapter 2: Ranking-based clustering.** For link-based clustering of heterogeneous information networks, we need to explore links across heterogeneous types of data. Recent studies develop a ranking-based clustering approach (e.g., RankClus [66] and NetClus [69]) that generates both clustering and ranking results efficiently. This approach is based on the observation that ranking and clustering can mutually enhance each other because objects highly ranked in each cluster may contribute more towards unambiguous clustering, and objects more dedicated to a cluster will be more likely to be highly ranked in the same cluster.

- **Chapter 3: Classification of heterogeneous information networks.** Classification can also take advantage of links in heterogeneous information networks. Knowledge can be effectively propagated across a heterogeneous network because the nodes of the same type that are linked similarly via the same typed links are likely to be similar. Moreover, following the idea of ranking-based clustering, one can explore ranking-based classification since objects highly ranked in a class are likely to play a more important role in classification. These ideas lead to effective algorithms, such as GNetMine [31] and RankClass [30].

In *Part II: Meta-Path-Based Similarity Search and Mining*, we introduce a systematic approach for dealing with general heterogeneous information networks with a specified network schema, using a meta-path-based methodology. Under this framework, similarity search and other mining tasks such as relationship prediction can be addressed by systematic exploration of the network meta structure.

- **Chapter 4: Meta-path-based similarity search.** Similarity search plays an important role in the analysis of networks. By considering different linkage paths (i.e., meta-path) in a network, one can derive various semantics on similarity in a heterogeneous information network. A meta-path-based similarity measure, PathSim, is introduced in [65], which aims at finding peer objects in the network. PathSim turns out to be more meaningful in many scenarios compared with random-walk-based similarity measures.

- **Chapter 5: Meta-path-based relationship prediction.** Heterogeneous information network brings interactions among multiple types of objects and hence the possibility of predicting relationships across heterogeneous typed objects. By systematically designing meta-path-based topological features and measures in the network, supervised models can be used to learn appropriate weights associated with different topological features in relationship prediction [62; 63].

In *Part III: Relation Strength-Aware Mining*, we address the issue that the heterogeneity of relations between object types often leads to different mining results that can be chosen by users. With user guidance, the strength of each relation can be automatically learned for improved mining.

- **Chapter 6: Relation strength-aware clustering with incomplete attributes.** By specifying a set of attributes, the strengths of different relations in heterogeneous information networks can be automatically learned to help network clustering [61].

- **Chapter 7: Integrating user-guided clustering with meta-path selection.** Different meta-paths in a heterogeneous information network represent different relations and carry different semantic meanings. User guidance, such as providing a small set of training examples for some object types, can indicate user preference on the clustering results. Then a preferred meta-path or a weighted meta-paths combination can be learned to achieve better consistency between mining results and the training examples [67].

PART I

Ranking-Based Clustering and Classification

CHAPTER 2

Ranking-Based Clustering

For link-based clustering of heterogeneous information networks, we need to explore links across heterogeneous types of data. In this chapter, we study how ranking can be computed for different types of objects using different types of links, and show how ranking and clustering mutually enhance each other and finally achieve reasonable ranking and clustering results. Two special cases of heterogeneous information networks, the bi-typed networks and the star networks, are studied.

2.1 OVERVIEW

A great many analytical techniques have been proposed toward a better understanding of information networks, though largely on homogeneous information networks, among which are two prominent ones: *ranking* and *clustering*. On one hand, *ranking* evaluates objects of information networks based on some *ranking function* that mathematically demonstrates characteristics of objects. With such functions, two objects can be compared, either qualitatively or quantitatively, in a partial order. PageRank [10] and HITS [34], among others, are perhaps the most well-known ranking algorithms over information networks. On the other hand, *clustering* groups objects based on a certain proximity measure so that similar objects are in the same cluster, whereas dissimilar ones are in different clusters. After all, as two fundamental analytical tools, ranking and clustering can be used to show the overall views of an information network, and hence have been be widely used in various applications.

Clustering and ranking are often regarded as *orthogonal* techniques, each applied independently to information network analysis. However, applying only one of them over an information network often leads to incomplete, or sometimes rather biased, analytical results. For instance, ranking objects over a whole information network without considering which clusters they belong to often leads to dumb results, e.g., ranking database and computer architecture venues or authors together may not make much sense; alternatively, clustering a large number of objects (e.g., thousands of authors) into one cluster without distinction is dull as well. However, integrating two functions together may lead to more comprehensible results, as shown in Example 2.1.

Example 2.1 (Ranking without/with clustering) Consider a set of venues from two areas of (1) DB/DM (i.e., *Database and Data Mining*) and HW/CA (i.e., *Hardware and Computer Architecture*), each having 10 venues, as shown in Table 2.1. We choose top 100 authors in each area from DBLP, according to their number of publications in the selected venues. With the authority ranking function specified in Section 2.2.1, our ranking-only algorithm gives top-10 ranked results in Table 2.2. Clearly, the results are rather dumb (because of the mixture of the areas) and are biased towards (i.e., ranked higher for) the HW/CA area. Moreover, such biased ranking result is caused not by the specific ranking function we chose, but by the inherent incomparability between the two areas.

Still consider the same dataset. If we cluster the venues in the DB/DM area and rank both venues and the authors relative to this cluster, the ranking results are shown in Table 2.3.

Table 2.1: A set of venues from two research areas	
DB/DM	{SIGMOD, VLDB, PODS, ICDE, ICDT, KDD, ICDM, CIKM, PAKDD, PKDD}
HW/CA	{ASPLOS, ISCA, DAC, MICRO, ICCAD, HPCA, ISLPED, CODES, DATE, VTS}

Table 2.2: Top-10 ranked venues and authors without clustering

Rank	Venue	Rank	Authors
1	DAC	1	Alberto L. Sangiovanni-Vincentelli
2	ICCAD	2	Robert K. Brayton
3	DATE	3	Massoud Pedram
4	ISLPED	4	Miodrag Potkonjak
5	VTS	5	Andrew B. Kahng
6	CODES	6	Kwang-Ting Cheng
7	ISCA	7	Lawrence T. Pileggi
8	VLDB	8	David Blaauw
9	SIGMOD	9	Jason Cong
10	ICDE	10	D. F. Wong

This example shows that good clustering indeed enhances ranking results. Furthermore, assigning ranks to objects often leads to better understanding of each cluster. By integrating both clustering and ranking, one can get more comprehensible results on networks.

In this chapter, we introduce two ranking-based clustering algorithms, RankClus and NetClus, for two special cases of heterogeneous information networks, namely *bi-typed networks* and *star networks*, respectively. For both cases, we need to use heterogeneous links to calculate ranking as well as ranking-derived clusters.

2.2 RANKCLUS

Let's first examine the task of clustering one type of objects (target objects) using other types of objects (attribute objects) and the links in the network. For example, given a bi-typed bibliographic network containing venues and authors, where links exist between venues and authors, and between authors and authors, we are interested in clustering venues into different clusters representing different research communities, using the authors and links in the network. In this section, we introduce an algorithm, RankClus, based on the bi-typed bibliographic network.

Table 2.3: Top-10 ranked venues and authors in DB/DM cluster

Rank	Venue	Rank	Authors
1	VLDB	1	H. V. Jagadish
2	SIGMOD	2	Surajit Chaudhuri
3	ICDE	3	Divesh Srivastava
4	PODS	4	Michael Stonebraker
5	KDD	5	Hector Garcia-Molina
6	CIKM	6	Jeffrey F. Naughton
7	ICDM	7	David J. DeWitt
8	PAKDD	8	Jiawei Han
9	ICDT	9	Rakesh Agrawal
10	PKDD	10	Raghu Ramakrishnan

Figure 2.1 illustrates a bi-typed bibliographic network, which contains two types of objects, venues (X) and authors (Y). Two types of links exist in this network: the author-venue publication links, with the weight indicating the number of papers an author has published in a venue, and the author-author co-authorship links, with the weight indicating the number of times two authors have collaborated. The bi-typed network can be represented by a block-wise adjacency matrix:

$$W = \begin{pmatrix} W_{XX} & W_{XY} \\ W_{YX} & W_{YY} \end{pmatrix},$$

where W_{XX}, W_{XY}, W_{YX}, and W_{YY} each denotes a type of relation between types of the subscripts. Formally, a bi-typed information network can be defined as follows.

Definition 2.2 (Bi-typed information network) Given two types of object sets X and Y, where $X = \{x_1, x_2, \ldots, x_m\}$, and $Y = \{y_1, y_2, \ldots, y_n\}$, the graph $G = (\mathcal{V}, \mathcal{E})$ is called a bi-typed information network on types X and Y, if $\mathcal{V} = X \cup Y$ and $\mathcal{E} \subseteq \mathcal{V} \times \mathcal{V}$.

The biggest issue in clustering target objects in a network is that unlike in traditional attribute-based dataset, the features for those objects are not explicitly given here. A straightforward way to generate clusters for target objects in a heterogeneous network is to first evaluate the similarity between target objects using a link-based approach, such as SimRank [28], and then apply graph clustering methods [44; 58] to generate clusters. However, to evaluate pair-wise similarity between objects in an information network is a space and time consuming task. Instead, RankClus explores rank distribution for each cluster to generate new measures for target objects, which are low-dimensional. The clusters are improved under the new measure space. More importantly, this measure can be further enhanced during the iterations of the algorithm, because better clustering leads to better ranking, and better ranking gives better ranking-based features thus better clustering results. That is, different from combining ranking and clustering in a two-stage procedure like facet ranking [16; 79], the quality of clustering and ranking can be mutually enhanced in RankClus.

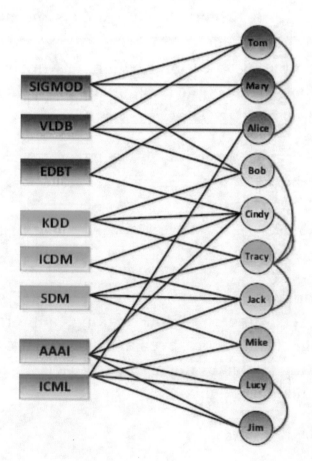

Figure 2.1: A bi-typed bibliographic network.

2.2.1 RANKING FUNCTIONS

Ranking function is critical in our ranking-based clustering algorithms, which not only provides rank scores for objects to distinguish their importance in a cluster, but also serves as a new feature extraction tool to improve the clustering quality. Current ranking functions are mostly defined on homogeneous networks, such as PageRank [10] and HITS [34]. In this section, we introduce two ranking functions based on the bi-typed bibliographic network: *Simple Ranking* and *Authority Ranking*. Ranking functions on more complex heterogeneous networks are discussed at the end of this section.

Simple Ranking

The simplest ranking of venues and authors is based on the number of publications, which is proportional to the numbers of papers accepted by a venue or published by an author.

Formally, given the bi-typed information network with types X and Y, and the adjacency matrix W, simple ranking generates the rank score of type X and type Y as follows:

$$
\begin{cases}
\vec{r}_X(x) = \dfrac{\sum_{j=1}^{n} W_{XY}(x, j)}{\sum_{i=1}^{m} \sum_{j=1}^{n} W_{XY}(i, j)} \\[2ex]
\vec{r}_Y(y) = \dfrac{\sum_{i=1}^{n} W_{XY}(i, y)}{\sum_{i=1}^{m} \sum_{j=1}^{n} W_{XY}(i, j)}
\end{cases}
\tag{2.1}
$$

The time complexity of Simple Ranking is $O(|\mathcal{E}|)$, where $|\mathcal{E}|$ is the number of links. According to simple ranking, authors publishing more papers will have higher rank score, even these papers are all in junk venues. In fact, simple ranking evaluates importance of each object according to the number of their immediate neighbors.

Authority Ranking

A more useful ranking function we propose here is *authority ranking*, which gives an object higher rank score if it has more authority. Ranking authority merely with publication information seems impossible at the first glance, as citation information could be unavailable or incomplete (such as in the DBLP data, where there is no citation information imported from Citeseer, ACM Digital Library, or Google Scholars). However, two simple empirical rules give us the first clues.

- Rule 1: Highly ranked authors publish *many* papers in highly ranked venues.

- Rule 2: Highly ranked venues attract *many* papers from highly ranked authors.

Note that these empirical rules are domain dependent and are usually given by the domain experts who know both the field and the dataset well. From the above heuristics, we define the iterative rank score formulas for authors and venues according to each other as follows.

According to Rule 1, each author's score is determined by the number of papers and their publication forums:

$$\vec{r}_Y(j) = \sum_{i=1}^{m} W_{YX}(j,i)\vec{r}_X(i) . \tag{2.2}$$

When author j publishes more papers, there are more nonzero and high weighted $W_{YX}(j,i)$, and when the author publishes papers in a higher ranked venue i, which means a higher $\vec{r}_X(i)$, the score of author j will be higher. At the end of each step, $\vec{r}_Y(j)$ is normalized by $\vec{r}_Y(j) \leftarrow \frac{\vec{r}_Y(j)}{\sum_{j'=1}^{n} \vec{r}_Y(j')}$.

According to Rule 2, the score of each venue is determined by the quantity and quality of papers in the venue, which is measured by their authors' rank scores,

$$\vec{r}_X(i) = \sum_{j=1}^{n} W_{XY}(i,j)\vec{r}_Y(j) . \tag{2.3}$$

When there are more papers appearing in venue i, there are more non-zero and high weighted $W_{XY}(i, j)$, and if the papers are published by a higher ranked author j, which means a higher $\vec{r}_Y(j)$, the score of venue i will be higher. The score vector is then normalized by $\vec{r}_X(i) \leftarrow \frac{\vec{r}_X(i)}{\sum_{i'=1}^{m} \vec{r}_X(i')}$.

Note that the normalization will not change the ranking position of an object, but it gives a relative importance score to each object. And as shown in RankClus [66], the two formulas will converge to the primary eigenvector of $W_{XY} W_{YX}$ and $W_{YX} W_{XY}$, respectively.

When considering the co-author information, the scoring function can be further refined by a third rule.

- Rule 3: The rank of an author is enhanced if he or she co-authors with many highly ranked authors.

Adding this new rule, we can calculate rank scores for authors by revising Equation (2.2) as

$$\vec{r}_Y(i) = \alpha \sum_{j=1}^{m} W_{YX}(i, j)\vec{r}_X(j) + (1 - \alpha) \sum_{j=1}^{n} W_{YY}(i, j)\vec{r}_Y(j), \qquad (2.4)$$

where parameter $\alpha \in [0, 1]$ determines how much weight to put on each factor, which can be assigned based on one's belief or learned by some training dataset.

Similarly, we can prove that $\vec{r}_Y$ should be the primary eigenvector of $\alpha W_{YX} W_{XY} + (1 - \alpha)W_{YY}$, and $\vec{r}_X$ should be the primary eigenvector of αW_{XY} $(I - (1 - \alpha)W_{YY})^{-1} W_{YX}$. Since the iterative process is a power method to calculate primary eigenvectors, the rank score will finally converge.

For authority ranking, the time complexity is $O(t|\mathcal{E}|)$, where t is the iteration number and $|\mathcal{E}|$ is the number of links in the graph. Note that, $|\mathcal{E}| = O(d|\mathcal{V}|) \ll |\mathcal{V}|^2$ in a sparse network, where $|\mathcal{V}|$ is the number of total objects in the network and d is the average link per object.

Different from simple ranking, authority ranking gives an importance measure to each object based on the whole network, rather than its immediate neighborhoods, by the score propagation over the whole network.

Alternative Ranking Functions

Although here we illustrate only two ranking functions, general ranking functions are not confined to them. In practice, a ranking function is not only related to the link property of an information network, but also depends on domain knowledge. For example, in many science fields, journals are given higher weights than conferences when evaluating an author. Moreover, although ranking functions in this section are defined on bi-typed networks, ranking function on heterogeneous networks with more types of objects can be similarly defined. For example, PopRank [51] is a possible framework for general heterogeneous network, which takes into account the impact both from the same type of objects and from the other types of objects, with different impact factors for different types. When ranking objects in information networks, junk or spam entities are often ranked higher than deserved. For example, authority ranking can be spammed by some bogus venues that accept any submissions due to their large numbers of accepted papers. Techniques that can best use expert knowledge, such as TrustRank [23], can be used to semi-automatically separate reputable, good objects from spam ones. Personalized PageRank [86], that can utilize expert ranking as query and generate rank distributions with respect to such knowledge, can be another choice to integrate expert knowledge.

2.2.2 FROM CONDITIONAL RANK DISTRIBUTIONS TO NEW CLUSTERING MEASURES

Given a bi-typed bibliographic network, suppose that we have an initial partition on target type X (venue type) and have calculated the conditional rank scores of venues and authors for each clustered network, the next issue becomes how to use the conditional rank scores to further improve the clustering results. Intuitively, for each venue cluster, which could form a research area, the rank scores of authors conditional to this cluster (or research area) should be distinct from that of the authors in other clusters.

This implies that these rank scores can be used to derive new features for objects for better clustering. Further, we treat these rank scores as from a discrete rank distribution, as they are non-negative values and summing up to 1, which indicates the subjective belief of how likely one may know an author or a venue according to their authority in each cluster.

Example 2.3 (Conditional rank distribution as cluster feature) Conditional rank distributions in different clusters are distinct from each other, especially when these clusters are reasonably well partitioned. Still using the network of the two-research-area example introduced in Section 2.1, we rank 200 authors based on two venue clusters, and the two conditional rank distributions are shown in Figure 2.2. From the figure, we can clearly see that DB/DM authors (with IDs from 1 to 100) rank high relative to the DB/DM venues, whereas rank extremely low relative to the HW/CA venues. A similar situation happens for the HW/CA authors (with IDs from 101 to 200).

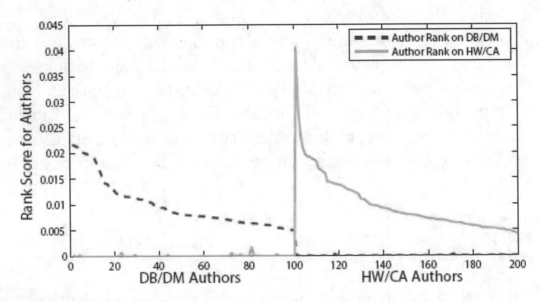

Figure 2.2: Authors' rank distributions over different clusters.

From Example 2.3, one can see that conditional rank distributions for attribute type in each cluster are rather different from each other, and can be used as measures to characterize each cluster. That is, for each cluster X_k, the conditional rank scores of X's and Y's, $\vec{r}_{X|X_k}$ and $\vec{r}_{Y|X_k}$, can be viewed as conditional rank distributions of X and Y, which in fact are the features for cluster X_k.

Cluster Membership for Each Target Object

Suppose we now know the clustering results for type X, which are X_1, $X_2, \ldots,$ and X_K, where K is the number of clusters. Also, according to some given ranking function, we have got conditional rank distribution over Y in each cluster X_k, which is $\vec{r}_{Y|X_k}$ ($k = 1, 2, \ldots, K$), and conditional rank distribution over X, which is $\vec{r}_{X|X_k}$ ($k = 1, 2, \ldots, K$). For simplicity, we use $p_k(Y)$ to denote $\vec{r}_{Y|X_k}$ and $p_k(X)$ to denote $\vec{r}_{X|X_k}$ in the following. We use $\pi_{i,k}$ to denote x_i's cluster membership for cluster k, which in fact is the posterior probability that x_i belongs to cluster k and satisfies $\sum_{k=1}^{K} \pi_{i,k} = 1$.

According to Bayes' rule, $p(k|x_i) \propto p(x_i|k)p(k)$. Since we already know $p(x_i|k)$, the conditional rank of x_i in cluster k, the goal is thus to estimate $p(k)$, the cluster size of k. In the DBLP scenario, the cluster of venue, e.g., the DB venues, can induce a subnetwork of venues and authors in that area. $p(k)$ can be considered as the proportion of papers belonging to the research area that is induced by the kth venue cluster, where each paper is represented by a link between a venue and an author. According to $p(k|x_i) \propto p(x_i|k)p(k)$, we can see that in general the higher its conditional rank in a cluster ($p(x_i|k)$), the higher possibility an object will belong to that cluster ($p(k|x_i)$). Since the conditional rank scores of X objects are propagated from the conditional rank scores of Y objects, we can also see that highly ranked attribute object has more impact on determining the cluster membership of a target object.

Example 2.4 (Cluster membership as object feature) Following Example 2.3, each venue x_i is represented as a two-dimensional cluster membership vector $(\pi_{i,1}, \pi_{i,2})$. We plot 20 venues according to their cluster membership vectors in Figure 2.3, where different styles of points represent different areas the venues really belong to. From the figure, we can see that the

DB/DM venues (denoted as *) and the HW/CA venues (denoted as +) are separated clearly under the new features in terms of cluster membership vectors, which are derived according to the conditional rank distributions of venues and authors with respective to the two research areas.

Parameter Estimation Using the EM Algorithm

In order to derive the cluster membership for each target object, we need to estimate the size proportion for each cluster $p(k)$ correctly, which can be viewed as the proportion of the links issued by the target objects belonging to cluster k. In our bi-typed bibliographic information network scenario, this is the proportion of papers belonging to the cluster.

We then build a mixture model for generating links issued by the target objects. Namely, each link between objects x_i and y_j is generated with the probability $p(x_i, y_j) = \sum_k p_k(x_i, y_j)p(k)$, where $p_k(x_i, y_j)$ denotes the probability of generating such a link in cluster k. We also make an independence assumption that an attribute object y_j issuing a link is independent to a target object x_i accepting this link, which is $p_k(x_i, y_j) = p_k(x_i)p_k(y_j)$. This assumption says once an author writes a paper, he is more likely to submit it to a highly ranked venue to improve his rank; while for venues, they are more likely to accept papers coming from highly ranked authors to improve its rank as well. This idea is similar to preferential attachment [4] of link formation for homogeneous networks, but we are considering more complex rank distributions instead of degrees of objects.

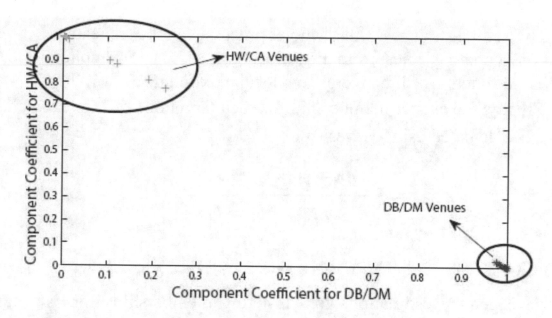

Figure 2.3: Venues' scatter plot based on 2-d cluster membership.

Let Θ be the K-dimensional parameter vector for $p(k)$'s. The likelihood of observing all the links between types X and Y under the parameter setting is then:

$$L(\Theta|W_{XY}) = p(W_{XY}|\Theta) = \prod_{i=1}^{m} \prod_{j=1}^{n} p(x_i, y_j|\Theta)^{W_{XY}(i,j)} , \qquad (2.5)$$

where $p(x_i, y_j|\Theta)$ is the probability to generate link $\langle x_i, y_j \rangle$, given current parameter Θ. The goal is to find the best Θ that maximizes the likelihood. We then apply the EM algorithm [8] to solve the problem. In the E-step, we calculate the conditional distribution $p(z = k|y_j, x_i, \Theta^0)$ based on the current value of Θ^0:

$$p(z = k|y_j, x_i, \Theta^0) \propto p(x_i, y_j|z = k)p(z = k|\Theta^0) = p_k(x_i)p_k(y_j)p^0(z = k) . \qquad (2.6)$$

In the M-Step, we update Θ according to the current Θ^0:

$$p(z = k) = \frac{\sum_{i=1}^{m} \sum_{j=1}^{n} W_{XY}(i, j)p(z = k|x_i, y_j, \Theta^0)}{\sum_{i=1}^{m} \sum_{j=1}^{n} W_{XY}(i, j)} . \qquad (2.7)$$

By setting $\Theta^0 = \Theta$, the whole process can be repeated. At each iteration, updating rules from Equations (2.6)–(2.7) are applied, and the likelihood function will converge to a local maximum.

Finally, the cluster membership for each target object x_i in each cluster k, $\pi_{i,k}$, is calculated using Bayes' rule:

$$\pi_{i,k} = p(z = k | x_i) = \frac{p_k(x_i)p(z = k)}{\sum_{l=1}^{K} p_l(x_i)p(z = l)}. \tag{2.8}$$

2.2.3 CLUSTER CENTERS AND DISTANCE MEASURE

After we get the estimations for clustering memberships for each target object x_i by evaluating mixture models, x_i can be represented as a K dimensional vector $\vec{s}_{xi} = (\pi_{i,1}, \pi_{i,2}, \ldots, \pi_{i,K})$. The centers for each cluster can thus be calculated accordingly, which is the mean of $\vec{s}_{xi}$ for all x_i in each cluster. Next, the distance between an object and cluster $D(x, X_k)$ is defined by 1 minus cosine similarity. The cluster label for each target object can be adjusted accordingly.

2.2.4 RANKCLUS: ALGORITHM SUMMARIZATION

To summarize, the input of RankClus is a bi-typed information network $G = \langle \{X \cup Y\}, W \rangle$, the ranking functions for X and Y, and the cluster number K. The output is K clusters of X with conditional rank scores for each x, and conditional rank scores for each y. The algorithm of RankClus is illustrated in Figure 2.4 and summarized in the following.

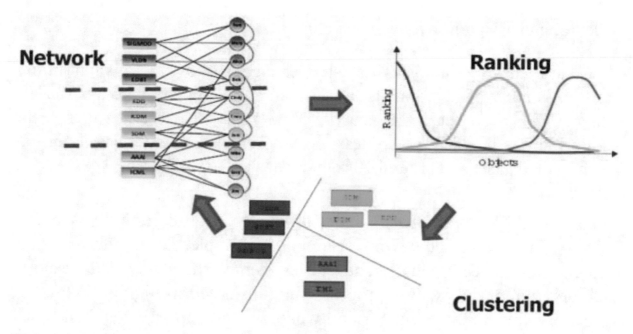

Figure 2.4: The illustration of the RankClus algorithm.

- Step 0: Initialization.
  The initial clusters for target objects are generated, by assigning each target object with a cluster label from 1 to K randomly.

- Step 1: Ranking for each cluster.
  Based on current clusters, K cluster-induced networks are generated accordingly, and the conditional rank distributions for types Y and X are calculated. In this step, we also need to judge whether any cluster is empty, which may be caused by the improper initialization of the algorithm. When some cluster is empty, the algorithm needs to restart in order to generate K clusters.

- Step 2: Estimation of the cluster membership vectors for target objects.
  In this step, we need to estimate the parameter Θ in the mixture model and get new representations for each target object and centers for each target cluster: $\vec{s}_x$ and $\vec{s}_{X_k}$. In practice, the number of iterations t for calculating Θ only needs to be set to a small number.

- Step 3: Cluster adjustment.
  In this step, we calculate the distance from each object to each cluster

center and assign it to the nearest cluster.

Repeat Steps 1, 2 and 3 until clusters change only by a very small ratio ε or the number of iterations is bigger than a predefined value *iterNum*. In practice, we can set $\varepsilon = 0$, and *iterNum* = 20. In our experiments, the algorithm will converge in less than 5 rounds in most cases for the synthetic dataset and around 10 rounds for the DBLP dataset.

Example 2.5 (Mutual improvement of clustering and ranking) We now apply our algorithm to the two-research-area example. The conditional rank distributions for each cluster and cluster memberships for each venue at each iteration of the running procedure are illustrated in Figure 2.5 (a)–(h). To better explain how our algorithm works, we set an extremely bad initial clustering as the initial state. In Cluster 1, there are 14 venues, half from the DB/DM area and half from the HW/CA area. Cluster 2 contains the remaining 6 venues, which are ICDT, CIKM, PKDD, ASPLOS, ISLPED, and CODES. We can see that the partition is quite unbalanced according to the size, and quite mixed according to the area. During the first iteration, the conditional rank distributions for two clusters are very similar to each other (Figure 2.5(a)), and venues are mixed up and biased to Cluster 2 (Figure 2.5(b)). However, we can still adjust their cluster labels according to the cluster centers, and most HW/CA venues go to Cluster 2 and most DB/DM venues go to Cluster 1. At the second iteration, conditional ranking improves somewhat (shown in Figure 2.5(c)) since the clustering (Figure 2.5(b)) is enhanced, and this time clustering results (Figure 2.5(d)) are enhanced dramatically, although they are still biased to one cluster (Cluster 1). At the third iteration, ranking results are improved significantly. Clusters and ranks are further adjusted afterwards, both of which are minor refinements.

At each iteration, the time complexity of RankClus is comprised of three parts: ranking part, mixture model estimation part, and clustering adjustment part. For ranking, if we use simple ranking, the time complexity is $O(|\mathcal{E}|)$. If we use authority ranking, the time complexity is $O(t_1|\mathcal{E}|)$, where $|\mathcal{E}|$ is the number of links, and t_1 is the number of iterations. For mixture model estimation, at each round, we need to calculate the conditional probability for each link in each cluster, the time complexity of which is $O(K|\mathcal{E}|)$. For clustering adjustment, we need to compute the distance between each object (m) and each cluster (K), and the dimension of each object is K, so the time complexity for this part is $O(mK^2)$. So, overall, the time complexity is $O(t(t_1|\mathcal{E}| + t_2(K|\mathcal{E}|) + mK^2))$, where t is the number of iterations of the whole algorithm and t_2 is the number of iterations of the mixture model. If the network is a sparse network, the time is almost linear to the number of objects.

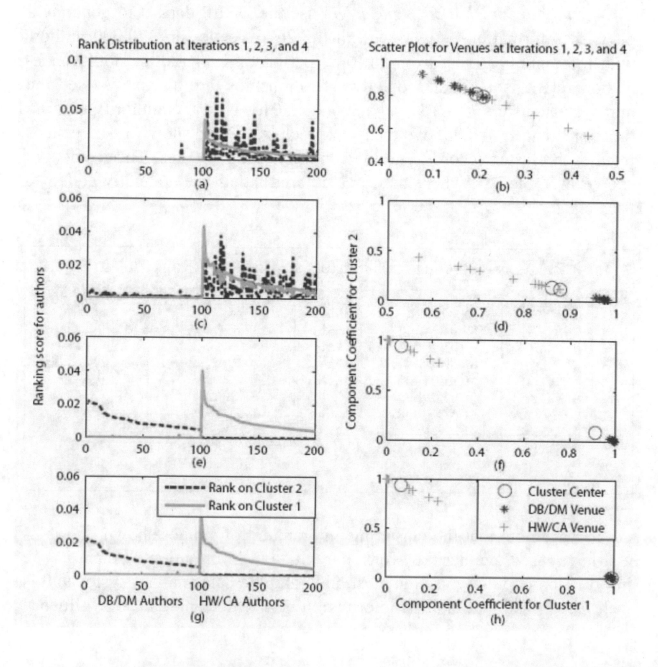

Figure 2.5: Mutual improvement of clusters and ranking through iterations.

2.2.5 EXPERIMENTAL RESULTS

We now show the effectiveness and efficiency of RankClus algorithm compared with other link-based algorithms, using both synthetic and real datasets.

Case Study on the DBLP Dataset We use the DBLP dataset to generate a bi-typed information network for all the 2676 venues and 20,000 authors with most publications, from the time period of 1998–2007. Both venue-author relationships and co-author relationships are used. We set the number of clusters $K = 15$, and apply RankClus with the authority ranking function, with $\alpha = 0.95$. We then pick 5 clusters, and show top-10 venues from each cluster according to the conditional rank scores. The results are shown in Table 2.4, where the research area labels are manually added to each cluster.

Table 2.4: Top-10 venues in 5 clusters generated by RankClus in DBLP

Rank	DB	Network	AI	Theory	IR
1	VLDB	INFOCOM	AAMAS	SODA	SIGIR
2	ICDE	SIGMETRICS	IJCAI	STOC	ACM Multimedia
3	SIGMOD	ICNP	AAAI	FOCS	CIKM
4	KDD	SIGCOMM	Agents	ICALP	TREC
5	ICDM	MOBICOM	AAAI/IAAI	CCC	JCDL
6	EDBT	ICDCS	ECAI	SPAA	CLEF
7	DASFAA	NETWORKING	RoboCup	PODC	WWW
8	PODS	MobiHoc	IAT	CRYPTO	ECDL
9	SSDBM	ISCC	ICMAS	APPROX-RANDOM	ECIR
10	SDM	SenSys	CP	EUROCRYPT	CIVR

Please note that the clustering and ranking of venues shown in Tables 2.4 have used neither keyword nor citation information, which is the information popularly used in most bibliographic data analysis systems. It is well recognized that citation information is crucial at judging the influence and impact of a venue or an author in a field. However, by exploring the publication entries only in the DBLP data, the RankClus algorithm can achieve comparable performance as citation studies for clustering and ranking venues and authors. This implies that the collection of publication entries without referring to the keyword and citation information can still tell a lot about the status of venues and authors in a scientific field.

Accuracy and Efficiency Study on Synthetic Data In order to compare accuracy among different clustering algorithms, we generate synthetic bi-typed information networks, which follow the properties of real information networks similar to DBLP. In our experiments, we first fixed the scale of the network and the distribution of links, but change configurations to adjust the density within each cluster and the separateness between different clusters, to obtain 5 different networks (Dataset1 to Dataset5). We set number of clusters $K = 3$, number of target objects in each cluster as $N_x = [10, 20, 15]$, and number of attribute objects in each cluster as $N_y = [500, 800, 700]$, which are the same for all the 5 datasets. Then we vary the number of links in each cluster (P) and the transition matrix of the proportion of links between different clusters (T), to get the following five datasets.

- Dataset1: medium separated and medium density.
  $P = [1000, 1500, 2000]$,
  $T = [0.8, 0.05, 0.15; 0.1, 0.8, 0.1; 0.1, 0.05, 0.85]$

- Dataset2: medium separated and low density.
  $P = [800, 1300, 1200]$,
  $T = [0.8, 0.05, 0.15; 0.1, 0.8, 0.1; 0.1, 0.05, 0.85]$

- Dataset3: medium separated and high density.
  $P = [2000, 3000, 4000]$,
  $T = [0.8, 0.05, 0.15; 0.1, 0.8, 0.1; 0.1, 0.05, 0.85]$

- Dataset4: highly separated and medium density.
  $P = [1000, 1500, 2000]$,
  $T = [0.9, 0.05, 0.05; 0.05, 0.9, 0.05; 0.1, 0.05, 0.85]$

- Dataset5: poorly separated and medium density.

  $P = [1000, 1500, 2000]$,

  $T = [0.7, 0.15, 0.15; 0.15, 0.7, 0.15; 0.15, 0.15, 0.7]$

We use the Normalized Mutual Information (NMI) [60] measure to compare the clustering accuracy among different algorithms. For N objects, K clusters, and two clustering results, let $n(i, j)$, $i, j = 1, 2, \ldots, K$, be the number of objects that has the cluster label i in the first clustering result (say generated by the algorithm) and cluster label j in the second clustering result (say the ground truth). We can then define joint distribution $p(i, j) = \frac{n(i, j)}{N}$, row distribution $p_1(j) = \sum_{i=1}^{K} p(i, j)$ and column distribution $p_2(i) = \sum_{j=1}^{K} p(i, j)$, and NMI is defined as:

$$\frac{\sum_{i=1}^{K} \sum_{j=1}^{K} p(i, j) \log(\frac{p(i,j)}{p_1(j)p_2(i)})}{\sqrt{\sum_{j=1}^{K} p_1(j) \log p_1(j) \sum_{i=1}^{K} p_2(i) \log p_2(i)}}. \tag{2.9}$$

We compare RankClus implemented with two ranking functions: Simple Ranking and Authority Ranking, with a state-of-the-art spectral clustering algorithm, the k-way N-cut algorithm [58], implemented with two link-based similarity functions, Jaccard Coefficient and SimRank [28].

Results for accuracy is summarized in Figure 2.6. From the results, we can see that, two versions of RankClus outperform in the first 4 datasets. RankClus with authority ranking is even better, since authority ranking gives a better rank distribution by utilizing the information of the whole network. Through the experiments, we observe that performance of two versions of RankClus and the N-Cut algorithm based on Jaccard coefficient are highly dependent on the data quality, in terms of cluster separateness and link density, while SimRank has a more stable performance, especially on the network that is sparse (Dataset5).

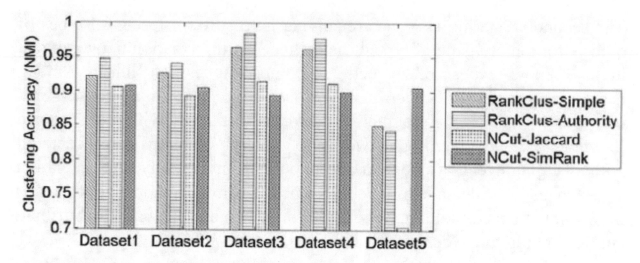

Figure 2.6: Accuracy comparison with baselines in terms of NMI. Dataset1: medium separated and medium density; Dataset2: medium separated and low density; Dataset3: medium separated and high density; Dataset4: highly separated and medium density; and Dataset5: poorly separated and medium density.

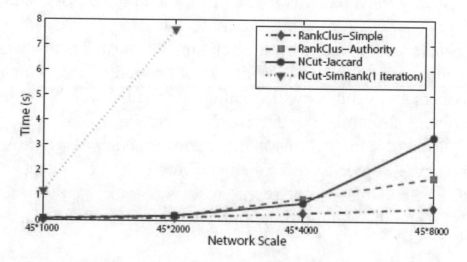

Figure 2.7: Efficiency comparison with baselines in terms of execution time.

Figure 2.7 summarizes the average execution time of different algorithms over four networks with different sizes. We can see that compared with the time-consuming SimRank algorithm, RankClus is very efficient and scalable.

2.3 NETCLUS

The second clustering task we are solving is to soft clustering all types of objects for a more general type of heterogeneous information networks that involve more types of objects and more types of links. Among heterogeneous networks, *networks with star network schema* (called *star networks*), such as bibliographic networks centered with papers (see Example 2.6) and tagging networks (e.g., http://delicious.com) centered with a tagging event, are popular and important. In fact, any *n*-nary relation set such as records in a relational database can be mapped into a star network, with each tuple in the relation as the center object and all attribute entities linking to the center object.

Example 2.6 (A star bibliographic information network) A bibliographic network contains rich information about research publications. It consists of nodes belonging to four types: paper (D), author (A), term (T), and venue (V). Semantically, each paper is *written* by a group of authors, *using* a set of terms, and *published* in a **venue** (a conference or a journal). Links exist between papers and authors by the relation of "write" and "written by," between papers and terms by the relation of "contain" and "contained in," between papers and venues by the relation of "publish" and "published by." The topological structure of a bibliographic network is shown in the left part of Figure 2.8, which forms a *star network schema*, where paper is a center type and all other types (called attribute types) of objects are linked via papers. The network can be represented as $G = (\mathcal{V}, \mathcal{E}, W)$, where $\mathcal{V} = A \cup V \cup T \cup D$, and the weight of the link $\langle x_i, x_j \rangle$, $w_{x_i x_j}$, is defined as:

$$
w_{x_i x_j} = \begin{cases} 1, & \text{if } x_i(x_j) \in A \cup V, \ x_j(x_i) \in D, \text{ and } x_i \text{ has link to } x_j, \\ c, & \text{if } x_i(x_j) \in T, \ x_j(x_i) \in D, \text{ and } x_i(x_j) \text{ appears } c \text{ times in } x_j(x_i), \\ 0, & \text{otherwise.} \end{cases}
$$

Definition 2.7 (Star network) An information network, $G = (\mathcal{V}, \mathcal{E}, W)$, with $T + 1$ types of objects (i.e., $\mathcal{V} = \{X_t\}_{t=0}^{T}$), is called with star network schema, if $\forall e = \langle x_i, x_j \rangle \in \mathcal{E}$, $x_i \in X_0 \wedge x_j \in X_t \ (t \neq 0)$, or vice versa. G is then called a **star network**. Type X_0 is the center type (called the **target type**) and X_t $(t \neq 0)$ are **attribute types**.

In contrast to traditional cluster definition, we propose NetClus to detect net-clusters that contain multiple types of objects and follow the schema of the original network, where each object can softly belong to multiple clusters. A net-cluster example is shown in Example 2.8.

Example 2.8 (The net-cluster of database area) A net-cluster of the *database area* consists of a set of database venues, authors, terms, and papers, and these objects belong to the database area with a (nontrivial) probability. Accordingly, we can present rank scores for attribute objects such as venues, authors and terms in its own type. With rank distribution, a user can easily grab the important objects in the area. Table 2.5 shows the top-ranked venues, authors, and terms in the area "*database*," generated from a 20-venue subnetwork from a "four-area" DBLP dataset (i.e., *database, data mining, information retrieval* and *artificial intelligence*) (see Section 2.3.5), using NetClus.

NetClus is designed for a heterogeneous network with the star network schema. It is a ranking-based iterative method following the idea of RankClus, that is, ranking is a good feature to help clustering. Different from RankClus, NetClus is able to deal with an arbitrary number of types of objects as long as the network is a star network, also the clusters generated are not groups of single typed objects but a set of subnetworks with the same topology as the input network. For a given star network and a specified number of clusters K, NetClus outputs K net-clusters (Figure 2.8). Each net-cluster is a sub-layer representing a concept of community of the network, which is an induced network from the clustered target objects, and attached with statistic information for each object in the network.

Table 2.5: Rank scores for venues, authors, and terms for the net-cluster of the database research area

Venue	Rank score	Author	Rank score	Term	Rank score
SIGMOD	0.315	Michael Stonebraker	0.0063	database	0.0529
VLDB	0.306	Surajit Chaudhuri	0.0057	system	0.0322
ICDE	0.194	C. Mohan	0.0053	query	0.0313
PODS	0.109	Michael J. Carey	0.0052	data	0.0251
EDBT	0.046	David J. DeWitt	0.0051	object	0.0138
CIKM	0.019	H. V. Jagadish	0.0043	management	0.0113
...	...	...	...	...	...

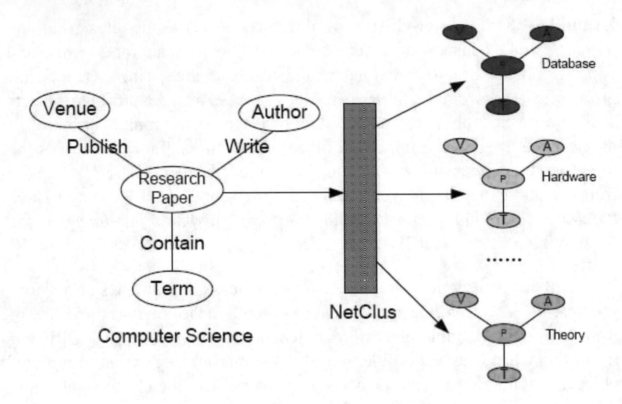

Figure 2.8: Illustration of clustering on a star bibliographic network into net-clusters.

Instead of generating pairwise similarities between objects, which is time consuming and difficult to define under a heterogeneous network, NetClus maps each target object, i.e., that from the center type, into a K-dimensional vector measure, where K is the number of clusters specified by the user. The probabilistic generative model for the target objects in each net-cluster is ranking-based, which factorizes a net-cluster into T independent components, where T is the number of attribute types. In this section, we use the star bibliographic network introduced in Example 2.6 to illustrate the NetClus algorithm.

2.3.1 RANKING FUNCTIONS

We have introduced ranking functions in Section 2.2.1, and now we re-examine the two ranking functions for the bibliographic network with a star network schema and illustrate some properties of the two ranking functions for a simple 3-(attribute-)typed star network.

Simple Ranking

Simple ranking is namely the simple occurrence counting for each object normalized in its own type. Given a network G, rank distribution for each attribute type of objects is defined as follows:

$$p(x|T_x, G) = \frac{\sum_{y \in N_G(x)} W_{xy}}{\sum_{x' \in T_x} \sum_{y \in N_G(x')} W_{x'y}}, \qquad (2.10)$$

where x is an object from type T_x, and $N_G(x)$ is the set of neighbors of x in G. For example, in the bibliographic network, the rank score for a venue using simple ranking will be proportional to the number of its published papers.

Authority Ranking

Authority ranking for each object is a ranking function that considers the authority propagation of objects in the whole network. Different from the bi-typed information network, we need to consider the rank score propagation over a path in a general heterogeneous information network. For a general star network G, the propagation of authority score from Type X to Type Y through the center type Z is defined as:

$$P(Y|T_Y, G) = W_{YZ} W_{ZX} P(X|T_X, G), \qquad (2.11)$$

where W_{YZ} and W_{ZX} are the weight matrices between the two corresponding types of objects, and can be normalized when necessary. Generally, authority score of one type of objects could be a combination of scores from different types of objects, e.g., that proposed in PopRank [51]. It turns out that the iteration method of calculating rank distribution is the power method to calculate the primary eigenvector of a square matrix denoting the strength between pairs of objects in that particular type, which can be achieved by selecting a walking path (or a combination of multiple paths) in the network. For more systematic definition of such paths, please refer to Chapter 4 for meta-path-based concepts.

In the DBLP dataset, according to the rules that (1) highly ranked venues accept many good papers published by highly ranked authors, and (2) highly ranked authors publish many good papers in highly ranked venues, we determine the iteration equation as:

$$\begin{aligned} P(V|T_V, G) &= W_{VD} D_{DA}^{-1} W_{DA} P(A|T_A, G) \\ P(A|T_A, G) &= W_{AD} D_{DV}^{-1} W_{DV} P(V|T_V, G), \end{aligned} \tag{2.12}$$

where D_{DA} and D_{DV} are the diagonal matrices with the diagonal value equaling to row sum of W_{DA} and W_{DV}, for the normalization purpose. The normalization simply means if a paper was written by multiple authors, we should consider the average rank score of these authors when calculating the rank score of a venue. Since all these matrices are sparse, in practice, the rank scores of objects need only be calculated iteratively according to their limited number of neighbors.

Integrating Ranking Functions with Prior Knowledge

In both ranking functions, prior distributions in different clusters for a certain type of objects can be integrated. For example, a user may give a few representative objects to serve as priors, like terms and venues in each research area. Priors for a given type X are represented in the form P_P $(X|T_X, k)$, $k = 1, 2, \ldots, K$. The prior is first propagated in the network in a Personalized PageRank [86] way, which propagates scores to objects that are not given in the priors. Then, the propagated prior is linearly combined with the rank distribution calculated by the given ranking function with parameter $\lambda_P \in [0, 1]$: the bigger the value, the more dependent on the prior is the final conditional rank distribution.

2.3.2 FRAMEWORK OF NETCLUS ALGORITHM

Here, we first introduce the general framework of NetClus, and each part of the algorithm will be explained in detail in the following sections. The general idea of the NetClus algorithm given the number of clusters K is composed of the following steps.

- Step 0: Generate initial partitions for target objects and induce initial net-clusters from the original network according to these partitions, i.e., $\{C_k^0\}_{k=1}^K$.

- Step 1: Build ranking-based probabilistic generative model for each net-cluster, i.e., $\{P(x|C_k^t)\}_{k=1}^K$.

- Step 2: Calculate the posterior probabilities for each target object $(p(C_k^t|x))$ and then adjust their cluster assignment according to the new measure defined by the posterior probabilities to each cluster.

- Step 3: Repeat Steps 1 and 2 until the cluster does not change significantly, i.e., $\{C_k^*\}_{k=1}^K \{C_k^t\}_{k=1}^K = \{C_k^{t-1}\}_{k=1}^K$.

- Step 4: Calculate the posterior probabilities for each attribute object $(p(C_k^*|x))$ in each net-cluster.

In all, the time complexity for NetClus is about linear to $|\mathcal{E}|$, the number of links in the network. When the network is very sparse, which is the real situation in most applications, the time complexity is almost linear to the number of objects in the network.

2.3.3 GENERATIVE MODEL FOR TARGET OBJECTS IN A NET-CLUSTER

According to many studies [4; 20; 50], preferential attachment and assortative mixing exist in many real-world networks, which means an object with a higher degree (i.e., high occurrences) has a higher probability to be attached with a link, and objects with higher occurrences tend to link more to each other. As in the DBLP dataset, 7.64% of the most productive authors publishes 74.2% of all the papers, among which 56.72% of papers are published in merely 8.62% of the biggest venues, which means large size venues and productive authors intend to co-appear via papers. We extend the heuristic by using rank score instead of degree of objects, which denotes the overall importance of an object in a network. Examples following this intuition include: webpage spammed by many low rank webpages linking to it (high degree but low rank) will not have too much chance to get a link from a real important webpage, and authors publishing many papers in junk venues will not increase his/her chance to publish a paper in highly ranked venues.

Under this observation, we simplify the network structure by proposing a probabilistic generative model for target objects, where a set of highly ranked attribute objects are more likely to co-appear to generate a center object. To explain this idea, we take the star bibliographic information network as a concrete example and show how the model works, where we assume the number of distinct objects in each type are $|A|$, $|V|$, $|T|$, and $|D|$, respectively, objects in each type are denoted as $A = \{a_1, a_2, \ldots, a_{|A|}\}$, $V = \{v_1, v_2, \ldots, v_{|V|}\}$, $T = \{t_1, t_2, \ldots, t_{|T|}\}$, and $D = \{d_1, d_2, \ldots, d_{|D|}\}$.

In order to simplify the complex network with multiple types of objects, we try to factorize the impact of different types of attribute objects and then model the generative behavior of target objects. The idea of factorizing a network is: we assume that given a network G, the probability to visit objects from different attribute types are independent to each other. Also, we make another independence assumption that within the same type of objects the probability to visit two different objects is independent to each other:

$$p(x_i, x_j | T_x, G) = p(x_i | T_x, G) \times p(x_j | T_x, G),$$

where $x_i, x_j \in T_x$ and T_x is some attribute type.

Now, we build the generative model for target objects given the rank distributions of attribute objects in the network G. Still using bibliographic network as an example, each paper d_i is written by several authors, published in one venue, and comprised of a bag of terms in the title. Therefore, a paper d_i is determined by several attribute objects, say x_{i1}, $x_{i2}, \ldots, x_{in_i}$, where n_i is the number of links d_i has. The probability to generate a paper d_i is equivalent to generating these attribute objects with the occurrence number indicated by the weight of the edge. Under the independency assumptions that we have made, the probability to generate a paper d_i in the network G is defined as follows:

$$p(d_i|G) = \prod_{x \in N_G(d_i)} p(x|T_x, G)^{W_{d_i, x}},$$

where $N_G(d_i)$ is the neighborhood of object d_i in network G, and T_x is used to denote the type of object x. Intuitively, a paper is generated in a cluster with high probability, if the venue it is published in, authors writing this paper and terms appeared in the title all have high probability in that cluster.

2.3.4 POSTERIOR PROBABILITY FOR TARGET OBJECTS AND ATTRIBUTE OBJECTS

Once we get the generative model for each net-cluster, we can calculate posterior probabilities for each target object. Now the problem becomes that suppose we know the generative probabilities for each target object generated from each cluster k, $k = 1, 2, \ldots, K$, what is the posterior probability that it is generated from cluster k? Here, K is the number of clusters given by the user. As some target objects may not belong to any of K net-cluster, we calculate $K + 1$ posterior probabilities for each target object instead of K, where the first K posterior probabilities are calculated for each real existing net-clusters $C_1, C_2, \ldots, C_K$, and the last one in fact is calculated for the original network G. Now, the generative model for target objects in G plays a role as a background model, and target objects that are not very related to any clusters will have high posterior probability in the background model. In this section, we will introduce the method to calculate posterior probabilities for both target objects and attribute objects.

According to the generative model for target objects, the generative probability for a target object d in the target type D in a subnetwork G_k is calculated according to the *conditional rank distributions* of attribute types in that subnetwork:

$$p(d|G_k) = \prod_{x \in N_{G_k}(d)} p(x|T_x, G_k)^{W_{d,x}}, \tag{2.13}$$

where $N_{G_k}(d)$ denotes the neighborhood of object d in subnetwork G_k. In Equation (2.13), in order to avoid zero probabilities in conditional rank scores, each conditional rank score should be first smoothed using global ranking:

$$P_S(X|T_X, G_k) = (1 - \lambda_S)P(X|T_X, G_k) + \lambda_S P(X|T_X, G), \tag{2.14}$$

where λ_S is a parameter that denotes how much we should utilize the rank distribution from the global ranking.

Smoothing [82] is a well-known technology in information retrieval. One of the reasons that smoothing is required in the language model is to deal with the zero probability problem for missing terms in a document. When calculating generative probabilities of target objects using our ranking-based generative model, we meet a similar problem. For example, for a paper in a given net-cluster, it may link to several objects whose rank score is zero in that cluster. If we simply assign the probability of the target object as zero in that cluster, we will miss the information provided by other objects. In fact, in initial rounds of clustering, objects may be assigned to wrong clusters, if we do not use smoothing technique, they may not have the chance to go back to the correct clusters.

Once a clustering is given on the input network G, say $C_1, C_2, \ldots, C_K$, we can calculate the posterior probability for each target object (say paper d_i) simply by Bayes' rule: $\pi_{i,k} \propto p(d_i|k) \times p(k)$, where $\pi_{i,k}$ is the probability that paper d_i is generated from cluster k given current generative model, and $p(k)$ denotes the relative size of cluster k, i.e., the probability that a paper belongs to cluster k overall, where $k = 1, 2, \ldots, K, K+1$.

In order to get the potential cluster size $p(k)$ for each cluster k, we choose cluster size $p(k)$ that maximizes log-likelihood to generate the whole collection of papers and then use the EM algorithm to get the local maximum for $p(k)$:

$$logL = \sum_{i=1}^{|D|} \log(p(d_i)) = \sum_{i=1}^{|D|} \log \left(\sum_{k=1}^{K+1} p(d_i|k)p(k) \right). \tag{2.15}$$

We use the EM algorithm to get $p(k)$ by simply using the following two iterative formulas, by initially setting $p^{(0)}(k) = \frac{1}{K+1}$:

$$\pi_{i,k}^{(t)} \propto p(d_i|k)p^{(t)}(k); \quad p^{(t+1)}(k) = \sum_{i=1}^{|D|} \pi_{i,k}^{(t)}/|D|.$$

When posterior probability is calculated for each target object in each cluster C_k, each target object d can be represented as a K dimensional vector: $\vec{v}(d_i) = (\pi_{i,1}, \pi_{i,2}, \ldots, \pi_{i,K})$. The center for each cluster C_k can be represented by the mean vector of all the target objects belonging to the cluster under the new measure. Next, we calculate cosine similarity between each target object and each center of cluster, and assign the target object into the cluster with the nearest center. A target object is now only belonging to one cluster, and we denote $p(k|d_i)$ as 1 if d_i is assigned to cluster k, 0 otherwise. A new subnetwork G_k can be induced by current target objects belonging to cluster k. The adjustment is an iterative process, until target objects do not change their cluster label significantly under the current measure. Note that, when measuring target objects, we do not use the posterior probability for background model. We make such choices with two reasons: first, the absolute value of posterior probability for background model should not affect the similarity between target objects; second, the sum of the first K posterior probabilities reflects the importance of an object in determining the cluster center.

The posterior probabilities for attribute objects $x \in A \cup V \cup T$ can be calculated as follows:

$$p(k|x) = \sum_{d \in N_G(x)} p(k, d|x) = \sum_{d \in N_G(x)} p(k|d)p(d|x) = \sum_{d \in N_G(x)} p(k|d)\frac{1}{|N_G(x)|}.$$

This simply implies that the probability of a venue belonging to cluster C_k is equal to the average posterior probabilities of papers published in the venue; similarly for authors and other attribute objects.

2.3.5 EXPERIMENTAL RESULTS

We now study the effectiveness of NetClus and compare it with several state-of-the-art baselines.

Dataset We build star bibliographic networks from DBLP according to Example 2.6. Two networks of different scales are studied. One is a big dataset ("all-area" dataset) which covers all the venues, authors, papers, and terms from DBLP. The other is a smaller dataset extracted from DBLP, containing 20 venues from four areas (hence called *"four-area"* dataset): *database, data mining, information retrieval*, and *artificial intelligence*. All authors have ever published papers on any of the 20 venues, and all these papers and terms appeared in the titles of these papers are included in the network. Using the "four-area" dataset, we are able to compare the clustering accuracy with several other methods.

Case Studies We first show the rank distributions in net-clusters we discovered using the "all-area" dataset, which is generated according to authority ranking for venues and authors, by setting venue type as priors and the cluster number as 8. We show four net-clusters in Table 2.6. Also, we can recursively apply NetClus to subnetworks derived from clusters and discover finer level net-clusters. Top-5 authors in a finer level net-cluster about XML area, which is derived from database subnetwork, are shown in Table 2.7.

Table 2.6: Top-5 venues in 4 net-clusters

Rank	DB and IS	Theory	AI	Software Engineering
1	SIGMOD	STOC	AAAI	ITC
2	VLDB	FOCS	UAI	VTS
3	ICDE	SIAM J. Comput.	IJCAI	POPL
4	SIGIR	SODA	Artif. Intell.	IEEE Trans. Computers
5	KDD	J. Comput. Syst. Sci.	NIPS	IEEE Design & Test of Compu.

Table 2.7: Top-5 authors in "XML" net-cluster

Rank	Author
1	Serge Abiteboul
2	Victor Vianu
3	Jerome Simeon
4	Michael J. Carey
5	Sophie Cluet

Study on Ranking Functions In Section 2.3.1, we proposed two ranking functions, namely simple ranking and authority ranking. Here, we study how low dimensional measure derived from rank distributions improve clustering and how clustering can improve this new measure in turn (Figure 2.9). In this study, term is always ranked by simple ranking, and venue and author are ranked by either authority ranking or simple ranking as two different settings.

First, we calculate average KL divergence between each conditional rank distribution and the global rank distribution for each attribute type X to measure the dissimilarity among different conditional rank distributions, which is denoted as $avgKL(X)$ for type X:

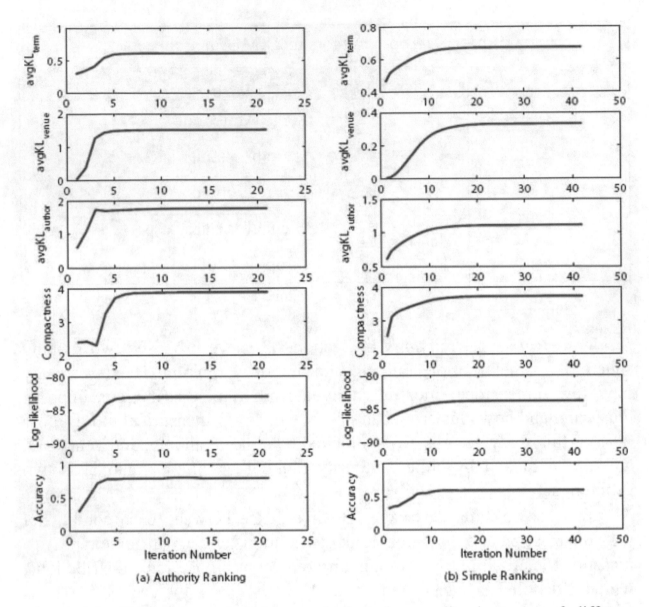

Figure 2.9: The change of ranking and clustering quality in terms of different measurements along with the iterations.

$$avg_{KL}(X) = \frac{1}{K} \sum_{k=1}^{K} D_{KL}(P(X|T_X, G_k)\|P(X|T_X, G)).$$

Second, in order to evaluate how good the new measure is generated in each round for clustering under the ranking function f, we use the compactness, denoted as C_f, which is defined as the average ratio between within-cluster similarity and between-cluster similarity using the new measure:

$$C_f = \frac{1}{|D|} \sum_{k=1}^{K} \sum_{i=1}^{|D_k|} \frac{s(d_{ki}, c_k)}{\sum_{k' \neq k} s(d_{ki}, c_{k'})/(K-1)} .$$

Third, we trace the accuracy of clustering results for target objects in each round of iteration, which is defined as:

$$accuracy = \frac{1}{|D|} \sum_{i=1}^{D} P_{true}(\cdot|d_i) \cdot P(\cdot|d_i) .$$

In other words, we calculate the percentage of papers that are assigned to the correct clusters. However, since $|D|$ is very large even in "four-area" dataset, we manually randomly labeled 100 papers into four clusters and use this paper set to calculate the accuracy.

Fourth, we trace the log-likelihood of the generative model along with the clustering iterations, which is defined in Equation (2.15). From Figure 2.9, we can see authority ranking is better than simple ranking in every measurement.

Accuracy Study In this section, we compare our algorithm with two other algorithms: the topic modeling algorithm PLSA [25] that merely uses term information for documents and RankClus that can only be applied to bi-typed networks. Since both of them cannot directly apply to heterogeneous networks with star network schema, we simplify the network when necessary. For PLSA, only the term type and paper type in the network are used, and we use the same term priors as in NetClus. The accuracy results for papers are in Table 2.8.

Table 2.8: Accuracy of paper clustering results using PLSA and NetClus		
	NetClus (A+V+T+D)	PLSA (T+D)
Accuracy	0.7705	0.608

Since RankClus can only cluster venues, we choose to measure the accuracy of venue cluster. For NetClus, cluster label is obtained according to the largest posterior probability, and Normalized Mutual Information (NMI) is used to measure the accuracy. Since the majority of the authors publish only a few papers, which may contain noise for correctly identifying the clustering of venues, we run RankClus algorithm by setting different thresholds to select subsets of authors. The results are shown in Table 2.9, where $d(a) > n$ means we select authors that have more than n publications to build the bi-typed bibliographic network. All the results are based on 20 runs.

Table 2.9: Accuracy of venue clustering results using RankClus and NetClus

	RankClus $d(a) > 0$	RankClus $d(a) > 5$	RankClus $d(a) > 10$	NetClus $d(a) > 0$
NMI	0.5232	0.8390	0.7573	**0.9753**

We can see that by using more types of objects in the network, NetClus performs much better than the two baselines that can only utilize partial information in the network.

CHAPTER 3

Classification of Heterogeneous Information Networks

Ming Ji, *Department of Computer Science, University of Illinois at Urbana-Champaign*

Classification can also take advantage of links in heterogeneous information networks. Knowledge can be effectively propagated across a heterogeneous network because the nodes of the same type that are linked similarly via the same typed links are likely to be similar. Moreover, following the idea of ranking-based clustering, one can explore ranking-based classification since objects highly ranked in a class are likely to play a more important role in classification. In this chapter, we show that by distinguishing the types of links in the networks during class label propagation, the classification accuracy can be significantly enhanced.

3.1 OVERVIEW

In many real-world applications, label information is available for some objects in a heterogeneous information network. Learning from such labeled and unlabeled data via transductive classification can lead to good knowledge extraction of the hidden network structure. Although classification on homogeneous networks has been studied for decades, classification on heterogeneous networks has not been explored until recently. Moreover, both classification and ranking of the nodes (or data objects) in such networks are essential for network analysis. But so far these approaches have generally been performed separately.

We consider the transductive classification problem on heterogeneous networked data objects which share a common topic. Only some objects in the given network are labeled, and we aim to predict labels for all types of the remaining objects. Besides, we combine ranking and classification in order to perform more accurate analysis of a heterogeneous information network. The intuition is that highly ranked objects within a class should play more important roles in classification. On the other hand, class membership information is important for determining a high quality ranking over a dataset. We believe it is therefore beneficial to integrate classification and ranking in a simultaneous, mutually enhancing process.

In this chapter, we first introduce GNetMine, a transductive classification framework on heterogeneous information networks. Then we introduce RankClass, a novel ranking-based classification framework based on GNetMine.

3.2 GNETMINE

As discussed before, sometimes, label information is available for some data objects in a heterogeneous information network. Learning from labeled and unlabeled data is often called semi-supervised learning [7; 85; 89], which aims to classify the unlabeled data based on known information. Classification can help discover the hidden structure of the information network, and give deep insight into understanding different roles played by each object. In fact, applications like research community discovery, fraud detection and product recommendation can all be cast as a classification problem [46; 56]. Generally, classification can be categorized into two groups: (1) *transductive classification* [45; 46; 77; 85; 89]: to predict labels for the given unlabeled data; and (2) *inductive classification* [7; 43; 48; 56; 70]: to construct a decision function in the whole data space. In this chapter, we focus on transductive classification, which is a common scenario in networked data.

Current studies about transductive classification on networked data [43; 45; 46; 56] mainly focus on homogeneous information networks, that is, networks composed of a single type of objects, as mentioned above. But in real life, there could be multiple types of objects which form heterogeneous information networks. As a natural generalization of classification on homogeneous networked data, we consider the problem of classifying heterogeneous networked data into classes, each of which is composed of multi-typed data sharing a common topic. For instance, a research community in a bibliographic information network contains not only authors, but also papers, venues and terms all belonging to the same research area. Other examples include movie networks in which movies, directors, actors and keywords relate to the same genre, and e-commerce networks where sellers, customers, items and tags belong to the same shopping category.

The general problem of classification has been well studied in the literature. Transductive classification on strongly-typed heterogeneous information networks, however, is much more challenging due to the following characteristics of data.

1. **Complexity of the network structure.** When dealing with the multi-typed network structure in a heterogeneous information network, one common solution is to transform it into a homogenous network and apply traditional classification methods [46; 56]. However, this simple transformation has several drawbacks. For instance, suppose we want to classify *papers* into different research areas. Existing methods would most likely extract a *citation* network from the whole bibliographic network. Then some valuable discriminative information is likely to be lost (e.g., *authors* of the paper, and the *venue* the paper is published in). Another solution to make use of the whole network is to ignore the type differences between objects and links. Nevertheless, different types of objects naturally have different data distributions, and different types of links have different semantic meanings, therefore treating them equally is unlikely to be optimal. It has been recognized [41; 66] that while mining heterogeneous information networks, the type differences among links and objects should be respected in order to generate more meaningful results.

2. **Lack of features.** Traditional classification methods usually learn from local features or attributes of the data. However, there is no natural

feature representation for all types of networked data. If we transform the link information into features, we will likely generate very high dimensional and sparse data as the number of objects increases. Moreover, even if we have feature representation for some objects in a heterogeneous information network, the features of different types of objects are in different spaces and are hardly comparable. This is another reason why traditional feature-based methods including Support Vector Machines, Naïve Bayes and logistic regression are difficult to apply in heterogeneous information networks.

3. **Lack of labels.** Many classification approaches need a reasonable amount of training examples. However, labels are expensive in many real applications. In a heterogeneous information network, we may even not be able to have a fully labeled subset of all types of objects for training. Label information for some types of objects are easy to obtain while labels for some other types are not. Therefore, a flexible transductive classifier should allow label propagation among different types of objects.

In this section, we introduce a graph-based regularization framework to address all three challenges, which simultaneously classifies all of the non-attributed, network-only data with an arbitrary network topology and number of object/link types, just based on the label information of any type(s) of objects and the link structure. By preserving consistency over each relation graph corresponding to each type of links separately, we explicitly respect the type differences in links and objects, thus encoding the typed information in a more organized way than traditional graph-based transductive classification on homogeneous networks.

3.2.1 THE CLASSIFICATION PROBLEM DEFINITION

In this section, we introduce several related concepts and notations, and then formally define the problem.

Given an arbitrary heterogeneous information network $G = (\mathcal{V}, \mathcal{E}, W)$, suppose it contains m types of data objects, denoted by $\mathcal{X}_1 = \{x_{11}, \ldots, x_{1n_1}\}, \ldots, \mathcal{X}_m = \{x_{m1}, \ldots, x_{mn_m}\}$, we have $\mathcal{V} = \bigcup_{i=1}^{m} \mathcal{X}_i$. Let $\mathcal{E}$ be the set of links between any two data objects of $\mathcal{V}$, and W is the set of weight values on the links. We use $W_{x_{ip}x_{jq}}$ denote the weight of the link between any two objects x_{ip} and x_{jq}, represented by $\langle x_{ip}, x_{jq} \rangle$.

A *class* in heterogeneous information network is similar to the concept of net-cluster in NetClus [69], which can be considered as a layer of the original network. Each node belongs to one class and the links are induced to the class of the network, if both nodes are from the same class. In other words, a class in a heterogeneous information network is actually a subnetwork containing multi-typed objects that are closely related to each other.

Now the problem can be formalized as follows.

Definition 3.1 (Transductive classification on heterogeneous information networks) Given a heterogeneous information network $G = ($ $\mathcal{V}, \mathcal{E}, W)$, a subset of data objects $\mathcal{V}' \subseteq \mathcal{V} = \bigcup_{i=1}^{m} \mathcal{X}_i$, which are labeled with values $\mathcal{Y}$ denoting which class each object belongs to, the goal is to predict the class labels for all the unlabeled objects $\mathcal{V} - \mathcal{V}'$.

We design a set of one-versus-all soft classifiers in the multi-class classification task. Suppose the number of classes is K. For any object type $\mathcal{X}_i$, $i \in \{1, \ldots, m\}$, we try to compute a class indicator matrix $\mathbf{F}_i = [f_i^{(1)}, \ldots, f_i^{(K)}] \in \mathbb{R}^{n_i \times K}$, where each $f_i^{(k)} = [f_{i1}^{(k)}, \ldots, f_{in_i}^{(k)}]^T$ measures the confidence that each object $x_{ip} \in \mathcal{X}_i$ belongs to class k. Then we can assign the p-th object in type $\mathcal{X}_i$ to class c_{ip} by finding the maximum value in the p-th row of $\mathbf{F}_i$: $c_{ip} = \arg\max_{1 \leq k \leq K} f_{ip}^{(k)}$.

In a heterogeneous information network, a relation graph G_{ij} can be built corresponding to each type of link relationships between two types of data objects $\mathcal{X}_i$ and $\mathcal{X}_j$, $i, j \in \{1, \ldots, m\}$. Note that it is possible for $i = j$. Let $\mathbf{R}_{ij}$ be an $n_i \times n_j$ relation matrix corresponding to graph G_{ij}. The element at the p-th row and q-th column of $\mathbf{R}_{ij}$ is denoted as $R_{ij,pq}$, representing weight on link $\langle x_{ip}, x_{jq} \rangle$. There are many ways to define the weights on the links, which can also incorporate domain knowledge. A simple definition is as follows:

$$R_{ij,pq} = \begin{cases} 1 & \text{if data objects } x_{ip} \text{ and } x_{jq} \text{ are linked together} \\ 0 & \text{otherwise.} \end{cases}$$

Here we consider undirected relation graphs such that $\mathbf{R}_{ij} = \mathbf{R}_{ji}^T$. In this way, each heterogeneous network G can be mathematically represented by a set of relation matrices $G = \{\mathbf{R}_{ij}\}_{i,j=1}^m$.

In order to encode label information, we basically set a vector $\mathbf{y}_i^{(k)} = [y_{i1}^{(k)}, \ldots, y_{in_i}^{(k)}]^T \in \mathbb{R}^{n_i}$ for each data object type $\mathcal{X}_i$ such that:

$$y_{ip}^{(k)} = \begin{cases} 1 & \text{if } x_{ip} \text{ is labeled to the } k\text{-th class} \\ 0 & \text{otherwise.} \end{cases}$$

Then for each class $k \in \{1, \ldots, K\}$, our goal is to infer a set of $f_i^{(k)}$ from $\mathbf{R}_{ij}$ and $\mathbf{y}_i^{(k)}$, $i, j \in \{1, \ldots, m\}$.

3.2.2 GRAPH-BASED REGULARIZATION FRAMEWORK

This section starts with a description of the intuition of the method, proceeds to the formulation of the problem using a graph-based regularization framework, and then introduces efficient computational schemes to solve the optimization problem.

Intuition

Consider a simple bibliographic information network in Figure 3.1. Four types of objects (*paper, author, venue* and *term*) are interconnected by multi-typed links (denoted by solid black lines). Suppose we want to classify them into research communities. Labeled objects are shaded, whereas the labels of unshaded objects are unknown. Given prior knowledge that author A_1, paper P_1 and venue V_1 belong to the area of data mining, it is easy to infer that author A_2 who *wrote* paper P_1, and term T_1 which is *contained* in P_1, are both highly related to data mining. Similarly, author A_3, venue V_2, and terms T_2 and T_3 are likely to belong to the database area, since they link directly to a database paper P_3. For paper P_2, things become more complicated because it is linked with both labeled and unlabeled objects. The confidence of belonging to a certain class may be transferred not only from labeled objects (venue V_1 and author A_4), but also from unlabeled ones (authors A_2 and A_3, terms T_1, T_2 and T_3). The classification process can be intuitively viewed as a process of knowledge propagation throughout the network as shown in Figure 3.1, where the thick shaded arrows indicate possible knowledge flow. The more links between an object x and other objects of class k, the higher the confidence that x belongs to class k. Accordingly, labeled objects serve as the source of prior knowledge. Although this intuition is essentially consistency preserving over the network, which is similar to [45] and [85], the interconnected relationships in heterogeneous information networks are more complex due to the typed information. *Knowledge propagation through different types of links contains different semantic meaning, and thus should be considered separately.*

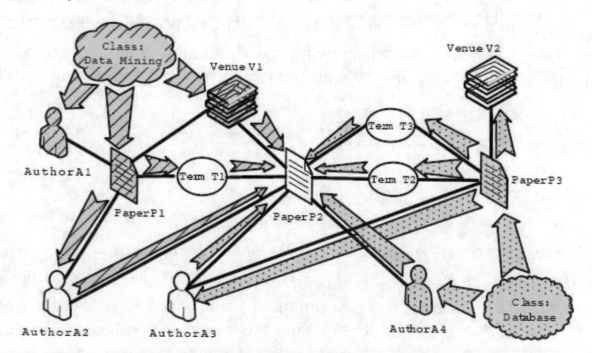

Figure 3.1: Knowledge propagation in a bibliographic information network.

In this way, the algorithm framework is based on the consistency assumption that the class assignments of two linked objects are likely to be similar. And the class prediction on labeled objects should be similar to their pre-assigned labels. In order to respect the type differences between links and objects, we ensure that such consistency is preserved over each relation graph corresponding to each type of links separately. The intuition could be formulated as follows.

1. The estimated confidence measure of two objects x_{ip} and x_{jq} belonging to class k, $f_{ip}^{(k)}$ and $f_{jq}^{(k)}$, should be similar if x_{ip} and x_{jq} are linked together, i.e., the weight value $R_{ij, pq} > 0$.

2. The confidence estimation $f_i^{(k)}$ should be similar to the ground truth, $y_i^{(k)}$.

The Algorithm

For each relation matrix $\mathbf{R}_{ij}$, we define a diagonal matrix $\mathbf{D}_{ij}$ of size $n_i \times n_i$. The (p, p)-th element of $\mathbf{D}_{ij}$ is the sum of the p-th row of $\mathbf{R}_{ij}$. Following the above discussion, $f_i^{(k)}$ should be as consistent as possible with the link information and prior knowledge within each relation graph, so we try to minimize the following objective function:

$$
J(f_1^{(k)}, \ldots, f_m^{(k)}) = \sum_{i,j=1}^{m} \lambda_{ij} \sum_{p=1}^{n_i} \sum_{q=1}^{n_j} R_{ij,pq} \left(\frac{1}{\sqrt{D_{ij,pp}}} f_{ip}^{(k)} - \frac{1}{\sqrt{D_{ji,qq}}} f_{jq}^{(k)} \right)^2
$$
$$
+ \sum_{i=1}^{m} \alpha_i (f_i^{(k)} - y_i^{(k)})^T (f_i^{(k)} - y_i^{(k)}). \tag{3.1}
$$

where $D_{ij,pp}$ is the (p, p)-th element of $\mathbf{D}_{ij}$, and $D_{ji,qq}$ is the (q, q)-th element of $\mathbf{D}_{ji}$. The first term in the objective function (3.1) is the *smoothness* constraints formulating the first intuition. This term is normalized by $\sqrt{D_{ij,pp}}$ and $\sqrt{D_{ji,qq}}$ in order to reduce the impact of popularity of nodes. In other words, we can, to some extent, suppress popular nodes from dominating the confidence estimations. The normalization technique is adopted in traditional graph-based learning and its effectiveness is well proved [85]. The second term minimizes the difference between the prediction results and the labels, reflecting the second intuition.

The trade-off among different terms is controlled by regularization parameters λ_{ij} and α_i, where $0 \leq \lambda_{ij} < 1$, $0 < \alpha_i < 1$. For $\forall i, j \in \{1, \ldots, m\}$, $\lambda_{ij} > 0$ indicates that object types $\mathcal{X}_i$ and $\mathcal{X}_j$ are linked together and this relationship is taken into consideration. The larger λ_{ij}, the more value is placed on the relationship between object types $\mathcal{X}_i$ and $\mathcal{X}_j$. For example, in a bibliographic information network, if a user believes that the links between *authors* and *papers* are more trustworthy and influential than the links between *venues* and *papers*, then the λ_{ij} corresponding to the *author-paper* relationship should be set larger than that of *venue-paper*, and the classification results will rely more on the *author-paper* relationship. Similarly, the value of α_i, to some extent, measures how much the user trusts the labels of object type $\mathcal{X}_i$. Similar strategy has been adopted in [41] to control the weights between different types of relations and objects.

To facilitate algorithm derivation, we define the normalized form of $\mathbf{R}_{ij}$:

$$\mathbf{S}_{ij} = \mathbf{D}_{ij}^{(-1/2)} \mathbf{R}_{ij} \mathbf{D}_{ji}^{(-1/2)}, i, j \in \{1, \ldots, m\} . \tag{3.2}$$

With simple algebraic formulations, the first term of (3.1) can be rewritten as:

$$\sum_{i,j=1}^{m} \lambda_{ij} \sum_{p=1}^{n_i} \sum_{q=1}^{n_j} R_{ij,pq} \left(\frac{1}{\sqrt{D_{ij,pp}}} f_{ip}^{(k)} - \frac{1}{\sqrt{D_{ji,qq}}} f_{jq}^{(k)} \right)^2$$

$$= \sum_{i,j=1}^{m} \lambda_{ij} \left((f_i^{(k)})^T f_i^{(k)} + (f_j^{(k)})^T f_j^{(k)} - 2(f_i^{(k)})^T \mathbf{S}_{ij} f_j^{(k)} \right) . \tag{3.3}$$

Then we can rewrite (3.1) in the following form:

$$J(f_1^{(k)}, \dots, f_m^{(k)}) = \sum_{i,j=1}^m \lambda_{ij} \left((f_i^{(k)})^T f_i^{(k)} + (f_j^{(k)})^T f_j^{(k)} - 2(f_i^{(k)})^T S_{ij} f_j^{(k)} \right)$$

$$+ \sum_{i=1}^m \alpha_i (f_i^{(k)} - y_i^{(k)})^T (f_i^{(k)} - y_i^{(k)}). \tag{3.4}$$

Closed Form and Iterative Solution

We first define $L_{ii} = I_i - S_{ii}$, where I_i is the identity matrix of size $n_i \times n_i$. Note that L_{ii} is the *normalized graph Laplacian* [13] of the homogeneous subnetwork on object type $\mathcal{X}_i$. We can show that the closed form solution is given by solving the following linear equation system:

$$f_i^{(k)} = \left((\sum_{j=1, j\neq i}^m \lambda_{ij} + \alpha_i) I_i + 2\lambda_{ii} L_{ii} \right)^{-1} \left(\alpha_i y_i^{(k)} + \sum_{j=1, j\neq i}^m \lambda_{ij} S_{ij} f_j^{(k)} \right), i \in \{1, \dots, m\}.$$

It can be proven that $\left((\sum_{j=1, j\neq i}^m \lambda_{ij} + \alpha_i) I_i + 2\lambda_{ii} L_{ii} \right)$ is positive definite and invertible.

Though the closed form solution is obtained, sometimes the iterative solution is preferable. We derive the iterative form of this algorithm as follows:

- Step 0: For $\forall k \in \{1, \dots, K\}$, $\forall i \in \{1, \dots, m\}$, initialize confidence estimates $f_i^{(k)}(0) = y_i^{(k)}$ and $t = 0$.

- Step 1: Based on the current $f_i^{(k)}(t)$, compute:

$$f_i^{(k)}(t+1) = \frac{\sum_{j=1, j\neq i}^m \lambda_{ij} S_{ij} f_j^{(k)}(t) + 2\lambda_{ii} S_{ii} f_i^{(k)}(t) + \alpha_i y_i^{(k)}}{\sum_{j=1, j\neq i}^m \lambda_{ij} + 2\lambda_{ii} + \alpha_i},$$

  for $\forall k \in \{1, \dots, K\}$, $\forall i \in \{1, \dots, m\}$.

- Step 2: Repeat step 1 with $t = t + 1$ until convergence, i.e., until $f_i^{(k)*} = f_i^{(k)}(t)$ do not change much for all i.

- Step 3: For each $i \in \{1, \ldots, m\}$, assign the class label to the p-th object of type $\mathcal{X}_i$ as $c_{ip} = arg\ max_{1 \le k \le K} f_{ip}^{(k)*}$, where $f_i^{(k)*} = [f_{i1}^{(k)*}, \ldots, f_{in_i}^{(k)*}]^T$.

Following an analysis similar to [85], the iterative algorithm can be proven to converge to the closed form solution. The iterative solution can be viewed as a natural extension of [85], where each object iteratively spreads label information to its neighbors until a global stable state is achieved. At the same time, we explicitly distinguish the semantic differences between the multi-typed links and objects by employing different normalized relation graphs corresponding to each type of links separately rather than a single graph covering all the instances.

The computational complexity of the iterative solution is $O(t\,K(|\mathcal{E}| + |\mathcal{V}|))$, where t is the number of iterations, $|\mathcal{E}|$ is the number of links in the network, and $|\mathcal{V}|$ is the number of total objects in the network.

The time complexity of the closed form solution is dependent on the particular network structure. In general, the iterative solution is more computationally efficient because it bypasses the matrix inversion operation.

3.3 RANKCLASS

Classification and *ranking* are two of the most fundamental analytical techniques. When label information is available for some of the data objects, *classification* makes use of the labeled data as well as the network structure to predict the class membership of the unlabeled data [46; 56]. On the other hand, *ranking* gives a partial ordering to objects in the network by evaluating the node/link properties using some ranking scheme, such as PageRank [10] or HITS [34]. Both classification and ranking have been widely studied and found to be applicable in a wide range of problems.

Traditionally, classification and ranking are regarded as orthogonal approaches, computed independently. However, adhering to such a strict dichotomy has serious downsides. As a concrete example, suppose we wish to classify the venues in Table 3.1 into two research areas. We wish to minimize the chance that the top venues are misclassified, not only to improve the classification results overall, but also because misclassifying a top venue is very likely to increase errors on many other objects that link to that venue, and are therefore greatly influenced by its label. We would thus like to more heavily penalize classification mistakes made on highly ranked venues, relative to a workshop of little influence. Providing a ranking of all venues within a research area can give users a clearer understanding of that field, rather than simply grouping venues into classes without noting their relative importance. On the other hand, the class membership of each venue is very valuable for characterizing that venue. Ranking all venues globally without considering any class information can often lead to meaningless results and apples-to-oranges comparisons. For instance, ranking database and information retrieval venues together may not make much sense since the top venues in these two fields cannot be reasonably compared, as shown in the second column of Table 3.2. These kinds of nonsensical ranking results are not caused by the specific ranking approach, but are rather due to the inherent incomparability between the two classes of venues. Thus, we suppose that combining classification with ranking may generate more informative results. The third and fourth columns in Table 3.2 illustrate this combined approach, showing the more meaningful venue ranking within each class.

In this study, we introduce RankClass, a new framework that groups objects into several prespecified classes, while generating the ranking information for each type of object within each class simultaneously in a heterogeneous information network. More accurate classification of objects increases the quality of the ranking within each class, since there is a higher guarantee that the ranking algorithm used will be comparing only objects of the same class. On the other hand, better ranking scores improve the performance of the classifier, by correctly identifying which objects are more important, and should therefore have a higher influence on the classifier's decisions. We use the ranking distribution of objects to characterize each class, and we treat each object's label information as a prior. By building a graph-based ranking model, different types of objects are ranked simultaneously within each class. Based on these ranking results, we estimate the relative importance or visibility of different parts of the network with regard to each class. In order to generate better within-class ranking, the network structure employed by the ranking model is adjusted so that the subnetwork composed of objects ranked high in each specific class is emphasized, while the subnetwork of the rest of the class is gradually weakened. Thus, as the network structure of each class becomes clearer, the ranking quality improves. Finally, the posterior probability of each object belonging to each class is estimated to determine each object's optimal class membership. Instead of performing ranking after classification, as facet ranking does [16; 79], RankClass essentially integrates ranking and classification, allowing both approaches to mutually enhance each other. RankClass iterates over this process until converging to a stable state. Experimental results show that RankClass both boosts the overall classification accuracy and constructs within-class rankings, which may be interpreted as meaningful summaries of each class.

Table 3.1: Venues from two research areas	
Database	SIGMOD, VLDB, ICDE, EDBT, PODS, . . .
Information Retrieval	SIGIR, ECIR, CIKM, WWW, WSDM, . . .

Table 3.2: Top-5 ranked venues in different settings.

Rank	Global Ranking	Within DB	Within IR
1	VLDB	VLDB	SIGIR
2	SIGIR	SIGMOD	ECIR
3	SIGMOD	ICDE	WWW
4	ICDE	PODS	CIKM
5	ECIR	EDBT	WSDM

Note that we still work on classification in the transductive setting. For convenience, we use $\mathcal{X}_i$ to denote both the set of objects belonging to the i-th type and the type name. In addition to grouping multi-typed objects into the pre-specified K classes, here we also aim to generate the ranking distribution of objects within each class k, which can be denoted as $P(x|T(x), k)$, $k = 1, \ldots, K \cdot T(x)$ denotes the type of object x. Note that different types of objects cannot be compared in a ranking. For example, it is not meaningful to create a ranking of venues and authors together in a bibliographic information network. Therefore, each ranking distribution is restricted to a single object type, i.e., $\sum_{p=1}^{n_i} P(x_{ip}|\mathcal{X}_i, k) = 1$.

Now the problem can be formalized as follows: given a heterogeneous information network $G = (\mathcal{V}, \mathcal{E}, W)$, a subset of data objects $\mathcal{V}' \subseteq \mathcal{V} = \bigcup_{i=1}^{m} \mathcal{X}_i$, which are labeled with values $\mathcal{Y}$ denoting which of the K pre-specified classes each object belongs to, predict the class labels for all the unlabeled objects $\mathcal{V} - \mathcal{V}'$ as well as the ranking distribution of objects within each class, $P(x|T(x),k)$, $x \in \mathcal{V}$, $k = 1, \ldots, K$.

The intuition behind RankClass is to build a graph-based ranking model that ranks multi-typed objects simultaneously, according to the relative importance of objects within each class. The initial ranking distribution of each class is determined by the labeled data. During each iteration, the ranking results are used to modify the network structure to allow the ranking model to generate higher quality within-class ranking.

3.3.1 THE FRAMEWORK OF RANKCLASS

We first introduce the general framework of RankClass. We will explain each part of the algorithm in detail in the following subsections.

- Step 0: Initialize the ranking distribution within each class according to the labeled data, i.e., $\{P(x|T(x), k)^0\}_{k=1}^{K}$. Initialize the set of network structures employed in the ranking model, $\{G_k^0\}_{k=1}^{K}$, as $G_k^0 = G$, $k = 1, \ldots, K$. Initialize $t = 1$.

- Step 1: Using the graph-based ranking model and the current set of network structures $\{G_k^{t-1}\}_{k=1}^{K}$, update the ranking distribution within each class k, i.e., $\{P(x|T(x), k)^t\}_{k=1}^{K}$.

- Step 2: Based on $\{P(x|T(x), k)^t\}_{k=1}^{K}$, adjust the network structure to favor within-class ranking, i.e., $\{G_k^t\}_{k=1}^{K}$.

- Step 3: Repeat steps 1 and 2, setting $t = t + 1$ until convergence, i.e., until $\{P(x|T(x), k)^*\}_{k=1}^K = \{P(x|T(x), k)^t\}_{k=1}^K$ do not change much for all $x \in \mathcal{V}$.

- Step 4: Based on $\{P(x|T(x), k)^*\}_{k=1}^K$, calculate the posterior probability for each object, i.e., $\{P(k|x, T(x))\}_{k=1}^K$. Assign the class label to object x as:

$$C(x) = \underset{1 \leq k \leq K}{\operatorname{argmax}} P(k|x, T(x))$$

When the number of classes K is given, the computational complexity is generally linear to the number of links and objects in the network.

3.3.2 GRAPH-BASED RANKING

Ranking is often used to evaluate the relative importance of objects in a collection. Here we rank objects within their own type and within a specific class. The higher an object x is ranked within class k, the more important x is for class k, and the more likely it is that x will be visited in class k. Clearly, within-class ranking is quite different from global ranking, and will vary throughout different classes.

The intuitive idea of the ranking scheme is authority propagation throughout the information network. Taking the bibliographic information network as an example, in a specific research area, it is natural to observe the following ranking rules that are similar to authority ranking introduced in Chapter 2.

1. Highly ranked venues publish many high-quality papers.

2. High-quality papers are often written by highly ranked authors.

3. High-quality papers often contain keywords that are highly representative of the papers' areas.

The above authority ranking rules can be generalized as follows: objects which are linked together in a network are more likely to share similar ranking scores. Therefore, the ranking of each object can be iteratively updated by looking at the rankings of its neighbors. The initial ranking distribution within a class k can be specified by the user. When data objects are labeled without ranking information in a general classification scenario, we can initialize the ranking as a uniform distribution over only the labeled data objects:

$$P(x_{ip}|\mathcal{X}_i, k)^0 = \begin{cases} 1/l_{ik} & \text{if } x_{ip} \text{ is labeled to class } k \\ 0 & \text{otherwise.} \end{cases}$$

where l_{ik} denotes the total number of objects of type $\mathcal{X}_i$ labeled to class k.

Suppose the current network structure used to estimate the ranking within class k is mathematically represented by the set of relation matrices: $G_k^{t-1} = \{\mathbf{R}_{ij}\}_{i,j=1}^m$. For each relation matrix $\mathbf{R}_{ij}$, we define a diagonal matrix $\mathbf{D}_{ij}$ of size $n_i \times n_i$. The (p, p)-th element of $\mathbf{D}_{ij}$ is the sum of the p-th row of $\mathbf{R}_{ij}$. Instead of using the original relation matrices in the authority propagation, we construct the normalized form of the relation matrices as follows:

$$\mathbf{S}_{ij} = \mathbf{D}_{ij}^{(-1/2)} \mathbf{R}_{ij} \mathbf{D}_{ji}^{(-1/2)}, i, j \in \{1, \ldots, m\}. \tag{3.5}$$

This normalization technique is adopted in traditional graph-based learning [85] in order to reduce the impact of node popularity. In other words, we can suppress popular nodes to some extent, to keep them from completely dominating the authority propagation. Notice that the normalization is applied separately to each relation matrix corresponding to each type of links, rather than to the whole network. In this way, the type differences between objects and links are well-preserved [31]. At the t-th iteration, the ranking distribution of object x_{ip} with regard to class k is updated as follows:

$$P(x_{ip}|\mathcal{X}_i, k)^t \propto \frac{\sum_{j=1}^m \lambda_{ij} \sum_{q=1}^{n_j} S_{ij,pq} P(x_{jq}|\mathcal{X}_j, k)^{t-1} + \alpha_i P(x_{ip}|\mathcal{X}_i, k)^0}{\sum_{j=1}^m \lambda_{ij} + \alpha_i}. \tag{3.6}$$

The first term of Equation (3.6) updates the ranking score of object x_{ip} by the summation of the ranking scores of its neighbors x_{jq}, weighted by the link strength $S_{ij,pq}$. The relative importance of neighbors of different types is controlled by $\lambda_{ij} \in [0, 1]$. The larger the value of λ_{ij}, the more value is placed on the relationship between object types $\mathcal{X}_i$ and $\mathcal{X}_j$. For example, in a bibliographic information network, if a user believes that the links between *authors* and *papers* are more trustworthy and influential than the links between *venues* and *papers*, then the λ_{ij} corresponding to the *author-paper* relationship should be set larger than that of *venue-paper*. As a result, the rank of a paper will rely more on the ranks of its authors than the rank of its publication venue. The parameters λ_{ij} can also be thought of as performing feature selection in the heterogeneous information network, that is, selecting which types of links are important in the ranking process.

The second term learns from the initial ranking distribution encoded in the labels, whose contribution is weighted by $\alpha_i \in [0, 1]$. A similar strategy has been adopted in [31; 41] to control the weights between different types of relations and objects. After each iteration, $P(x_{ip}|\mathcal{X}_i, k)^t$ is normalized such that $\sum_{p=1}^{n_i} P(x_{ip}|\mathcal{X}_i, k)^t = 1$, $\forall i = 1, \ldots, m$, $k = 1, \ldots, K$, in order to stay consistent with the mathematical definition of a ranking distribution.

We employ the authority propagation scheme in Equation (3.6) to estimate the ranking distribution instead of other simple measures computed according to the network topology (e.g., the degree of each object). This choice was made since we aim to rank objects with regard to each class by utilizing the current soft classification results. Therefore, if the ranking of an object were merely based on the network topology, it would be the same for all classes. By learning from the label information in the graph-based authority propagation method, the ranking of each object within different classes will be computed differently, which is more suitable for our setting.

Following a similar analysis to [31] and [86], the updating scheme in Equation (3.6) can be proven to converge to the closed form solution of minimizing the following objective function:

$$J(P(x_{ip}|\mathcal{X}_i, k))$$

$$= \sum_{i,j=1}^{m} \lambda_{ij} \sum_{p=1}^{n_i} \sum_{q=1}^{n_j} S_{ij,pq} \left(P(x_{ip}|\mathcal{X}_i, k) - P(x_{jq}|\mathcal{X}_j, k)\right)^2$$

$$+ \sum_{i=1}^{m} \alpha_i \sum_{p=1}^{n_i} (P(x_{ip}|\mathcal{X}_i, k) - P(x_{ip}|\mathcal{X}_i, k)^0)^2 , \tag{3.7}$$

which shares a similar theoretical foundation with GNetMine [31] that preserves consistency over each relation graph corresponding separately to each link type. However, we extend the graph-based regularization framework to rank objects within each class, which is conceptually different from GNetMine.

3.3.3 ADJUSTING THE NETWORK

Although graph-based ranking considers class information by incorporating the labeled data, it still ranks all object types in the global network. Instead, a within-class ranking should be performed over the subnetwork corresponding to each specific class. The cleaner the network structure, the higher the ranking quality. Therefore, the ranking within each class should be performed over a different subnetwork, rather than employing the same global network for every class. The network structure is mathematically represented by the weight values on the links. Thus, extracting the subnetwork belonging to class k is equivalent to increasing the weight on the links within the corresponding subnetwork, and decreasing the weight on the links in the rest of the network. It is straightforward to verify that multiplying $\mathbf{R}_{ij}$ by any positive constant c will not change the value of $\mathbf{S}_{ij}$. So increasing the weights on the links within a subnetwork should be performed relative to the weight on the links of other parts of the network. In other words, we can increase or decrease the absolute values of the weights on the links in the whole network, as long as the weights on the links of the subnetwork belonging to class k are larger than those on the links belonging to the rest of the network. Let $G_k^t = \{\mathbf{R}_{ij}^t(k)\}_{i,j=1}^{m}$. Here is a simple scheme that updates the network structure so as to favor the ranking within each class k, given the current ranking distribution $P(x|T(x), k)^t$:

$$R^t_{ij,pq}(k) = R_{ij,pq} \times \left(r(t) + \sqrt{\frac{P(x_{ip}|\mathcal{X}_i, k)^t}{\max_p P(x_{ip}|\mathcal{X}_i, k)^t} \frac{P(x_{jq}|\mathcal{X}_j, k)^t}{\max_q P(x_{jq}|\mathcal{X}_j, k)^t}} \right) . \qquad (3.8)$$

Recall that $\mathbf{R}_{ij}$ is the relation matrix corresponding to the links between object types $\mathcal{X}_i$ and $\mathcal{X}_j$ in the original network. Using the above updating scheme, the weight of each link $\langle x_{ip}, x_{jq} \rangle$ is increased in proportion to the geometric mean of the ranking scores of x_{ip} and x_{jq}, which are scaled to the interval of $[0, 1]$. The higher the rankings of x_{ip} and x_{jq}, the more important the link between them (i.e., $\langle x_{ip}, x_{jq} \rangle$) is in class k. The weight on that link should therefore be increased. Note that instead of creating hard partitions of the original network into classes, we simply increase the weights on the links that are important to classes k. This is because at any time in the iteration, the current classes represented by the ranking distributions are not very accurate, and the results will be more stable if we consider both the global network structure and the current ranking results. By gently increasing the weights of links in the subnetwork of class k, we gradually extract the correct subnetwork from the global network, since the weights of links in the rest of the network will decrease to very low values. Note that this adjustment of the network structure still respects the differences among the various types of objects and links.

In Equation (3.8), $r(t)$ is a positive parameter that does not allow the weights of links to drop to 0 in the first several iterations, when the authority scores have not propagated very far throughout the network and $P(x|T(x), k)^t$ are close to 0 in value for many objects. As discussed above, multiplying $\mathbf{R}_{ij}$ by any positive constant will not change the value of $\mathbf{S}_{ij}$. Therefore, it is essentially the ratio between $r(t)$ and $\sqrt{\frac{P(x_{ip}|k)^t}{\max_p P(x_{ip}|k)^t} \times \frac{P(x_{jq}|k)^t}{\max_q P(x_{jq}|k)^t}}$ that determines how much the original network structure and the current ranking distribution, respectively, contribute to the adjusted network G^t_k. Since we hope to progressively extract the subnetwork belonging to each class k, and we want to gradually reduce the weights of links that do not belong to class k down to 0, we decrease $r(t)$ exponentially by setting $r(t) = \frac{1}{2^t}$.

Equation (3.8) is not the only way to gradually increase the weights of links between highly ranked objects in class k. For instance, the geometric mean of $\frac{P(x_{ip}|k)^t}{\max_p P(x_{ip}|k)^t}$ and $\frac{P(x_{jq}|k)^t}{\max_q P(x_{jq}|k)^t}$ can be replaced by the arithmetic mean, and $r(t)$ can be any positive function that decreases with t. We show that even such simple adjustments as shown above can boost both the classification and ranking performance of RankClass.

3.3.4 POSTERIOR PROBABILITY CALCULATION

Once the ranking distribution of each class has been computed by the iterative algorithm, we can calculate the posterior probability of each object of type $\mathcal{X}_i$ belonging to class k simply by Bayes' rule:

$$P(k|x_{ip}, \mathcal{X}_i) \propto P(x_{ip}|\mathcal{X}_i, k)P(k|\mathcal{X}_i),$$

where $P(x_{ip}|\mathcal{X}_i, k) = P(x_{ip}|\mathcal{X}_i, k)^*$, and $P(k|\mathcal{X}_i)$ represents the relative size of class k among type $\mathcal{X}_i$, which should also be estimated. We choose the $P(k|\mathcal{X}_i)$ that maximizes the likelihood of generating the set of objects of type $\mathcal{X}_i$:

$$
\begin{aligned}
&\log L(x_{i1}, \ldots, x_{in_i}|\mathcal{X}_i) \\
&= \sum_{p=1}^{n_i} \log P(x_{ip}|\mathcal{X}_i) \\
&= \sum_{p=1}^{n_i} \log \left(\sum_{k=1}^{K} P(x_{ip}|\mathcal{X}_i, k) P(k|\mathcal{X}_i) \right).
\end{aligned}
\tag{3.9}
$$

By employing the EM algorithm, $P(k|\mathcal{X}_i)$ can be iteratively estimated using the following two equations:

$$P(k|x_{ip}, \mathcal{X}_i)^t \quad \propto \quad P(x_{ip}|\mathcal{X}_i, k)\, P(k|\mathcal{X}_i)^t$$

$$P(k|\mathcal{X}_i)^t \quad = \quad \sum_{p=1}^{n_i} P(k|x_{ip}, \mathcal{X}_i)^t / n_i\,,$$

where $P(k|\mathcal{X}_i)$ is initialized uniformly as $P(k|\mathcal{X}_i)^0 = 1/K$.

3.4 EXPERIMENTAL RESULTS

In this section, we apply the classification algorithm (denoted by GNetMine) and the ranking-based classification scheme (RankClass) introduced in this chapter, to the DBLP bibliographic network. We try to classify the bibliographic data into research communities, each of which consists of multi-typed objects closely related to the same area. The following five classification methods on information networks are compared:

- Ranking-based classification in heterogeneous information networks (RankClass);

- Graph-based regularization framework for transductive classification in heterogeneous information networks (GNetMine);

- Learning with Local and Global Consistency (LLGC) [85];

- Weighted-vote Relational Neighbor Classifier (wvRN) [45; 46];

- Network-only Link-based Classification (nLB) [46; 56].

LLGC is a graph-based transductive classification algorithm for homogeneous networks, while GNetMine is its extension, which works on heterogeneous information networks. Weighted-vote relational neighbor classifier and link-based classification are two popular classification methods for networked data. Since a feature representation of nodes is not available for our problem, we use the network-only derivative of the link-based classifier (nLB) [46], which creates a feature vector for each node based on neighboring information. Note that LLGC, wvRN, and nLB are classifiers which work with homogeneous networks, and cannot be directly applied to heterogeneous information networks. In order to compare all of the above algorithms, we can transform the heterogeneous DBLP network into a homogeneous network in two ways: (1) disregard the type differences between objects and treat all objects as the same type; or (2) extract a homogeneous subnetwork on one single type of object, if that object type is partially labeled. We try both approaches in the accuracy study.

3.4.1 DATASET

We use the same "four-area" dataset described in Chapter 2, which is a subnetwork of the DBLP network. As previously discussed, this heterogeneous information network is composed of four types of objects: paper, venue, author, and term. Among the four types of objects, we have three types of link relationships: paper-venue, paper-author, and paper-term.

We randomly choose a subset of labeled objects and use their label information in the learning process. The classification accuracy is evaluated by comparing with manually labeled results on the rest of the labeled objects. Since terms are difficult to label even manually, as many terms may belong to multiple areas, we do not evaluate the accuracy on terms here.

3.4.2 ACCURACY STUDY

In order to address the label scarcity problem in real life, we randomly choose $(a\%, p\%) = [(0.1\%, 0.1\%), (0.2\%, 0.2\%), \ldots, (0.5\%, 0.5\%)]$ of authors and papers, and use their label information in the classification task. For each $(a\%, p\%)$, we average the performance scores over 10 random selections of the labeled set. Note that the very small percentage of labeled objects here are likely to be disconnected, so we may not even be able to extract a fully labeled subnetwork for training, making many state-of-the-art algorithms inapplicable.

We set the parameters of LLGC to the optimal values, which were determined experimentally. For the introduced GNetMine and RankClass method, as discussed above, the parameters λ_{ij} are used to select which types of links are important in the classification and ranking processes, respectively. We consider all types of objects and links to be important in the DBLP network, so we set $\alpha_i = 0.1$, $\lambda_{ij} = 0.2$, $\forall i, j \in \{1, \ldots, m\}$. This may not be the optimal choice, but it is good enough to demonstrate the effectiveness of the introduced algorithms. Since labels are given for selected authors and papers, the results on venues of wvRN, nLB, and LLGC can only be obtained by mining the original heterogeneous information network (denoted by A-V-P-T) and disregarding the type differences between objects and links. While classifying authors and papers, we also tried constructing homogeneous *author-author* (A-A) and *paper-paper* (P-P) subnetworks in various ways, where the best results reported for authors are given by the co-author network, and the best results for papers are generated by linking two papers if they are published in the same venue. Note that there is no label information given for venues, so we cannot build a venue-venue (V-V) subnetwork for classification. We show the classification accuracy on authors, papers, and venues in Tables 3.3, 3.4 and 3.5, respectively. The last row of each table records the average classification accuracy while varying the percentage of labeled data.

Table 3.3: Comparison of classification accuracy on authors (%)

(a%, p%) of authors and papers labeled	nLB (A-A)	nLB (A-V-P-T)	wvRN (A-A)	wvRN (A-V-P-T)	LLGC (A-A)	LLGC (A-V-P-T)	GNetMine (A-V-P-T)	RankClass (A-V-P-T)
(0.1%, 0.1%)	25.4	26.0	40.8	34.1	41.4	61.3	82.9	85.4
(0.2%, 0.2%)	28.3	26.0	46.0	41.2	44.7	62.2	83.4	88.0
(0.3%, 0.3%)	28.4	27.4	48.6	42.5	48.8	65.7	86.7	88.5
(0.4%, 0.4%)	30.7	26.7	46.3	45.6	48.7	66.0	87.2	88.4
(0.5%, 0.5%)	29.8	27.3	49.0	51.4	50.6	68.9	87.5	89.2
average	28.5	26.7	46.3	43.0	46.8	64.8	85.5	87.9

Table 3.4: Comparison of classification accuracy on papers (%).

(a%, p%) of authors and papers labeled	nLB (P-P)	nLB (A-V-P-T)	wvRN (P-P)	wvRN (A-V-P-T)	LLGC (P-P)	LLGC (A-V-P-T)	GNetMine (A-V-P-T)	RankClass (A-V-P-T)
(0.1%, 0.1%)	49.8	31.5	62.0	42.0	67.2	62.7	79.2	77.7
(0.2%, 0.2%)	73.1	40.3	71.7	49.7	72.8	65.5	83.5	83.0
(0.3%, 0.3%)	77.9	35.4	77.9	54.3	76.8	66.6	83.2	83.6
(0.4%, 0.4%)	79.1	38.6	78.1	54.4	77.9	70.5	83.7	84.7
(0.5%, 0.5%)	80.7	39.3	77.9	53.5	79.0	73.5	84.1	84.8
average	72.1	37.0	73.5	50.8	74.7	67.8	82.7	82.8

Table 3.5: Comparison of classification accuracy on venues (%).

(a%, p%) of authors and papers labeled	nLB (A-V-P-T)	wvRN (A-V-P-T)	LLGC (A-V-P-T)	GNetMine (A-V-P-T)	RankClass (A-V-P-T)
(0.1%, 0.1%)	25.5	43.5	79.0	81.0	85.0
(0.2%, 0.2%)	22.5	56.0	83.5	85.0	85.5
(0.3%, 0.3%)	25.0	59.0	87.0	87.0	90.0
(0.4%, 0.4%)	25.0	57.0	86.5	89.5	92.0
(0.5%, 0.5%)	25.0	68.0	90.0	94.0	95.0
average	24.6	56.7	85.2	87.3	89.5

Table 3.6: Top-5 venues related to each research area generated by GNetMine and RankClass

GNetMine				RankClass			
Database	Data Mining	AI	IR	Database	Data Mining	AI	IR
VLDB	SDM	IJCAI	SIGIR	VLDB	KDD	IJCAI	SIGIR
ICDE	KDD	AAAI	ECIR	SIGMOD	SDM	AAAI	ECIR
SIGMOD	ICDM	ICML	CIKM	ICDE	ICDM	ICML	CIKM
PODS	PAKDD	CVPR	IJCAI	PODS	PKDD	CVPR	WWW
CIKM	PKDD	ECML	CVPR	EDBT	PAKDD	ECML	WSDM

Table 3.7: Top-5 terms related to each research area generated by GNetMine and RankClass. GNetMine RankClass

GNetMine				RankClass			
Database	Data Mining	AI	IR	Database	Data Mining	AI	IR
interlocking	rare	failing	helps	data	mining	learning	retrieval
deindexing	extreme	interleaved	specificity	database	data	knowledge	information
seed	scan	cognition	sponsored	query	clustering	reasoning	search
bitemporal	mining	literals	relevance	system	frequent	logic	web
debugging	associations	configuration	information	xml	classification	model	text

RankClass outperforms all other algorithms when classifying authors, papers, and venues. Note that even though the number of authors is much higher than the number of venues, RankClass achieves comparable accuracy for both of these types of objects. While classifying authors and papers, it is interesting to note that wvRN and nLB perform better on the *author-author* and *paper-paper* subnetworks than on the whole heterogeneous information network. We observe a similar result when we use LLGC to classify papers. These results serve to verify that homogeneous classifiers like wvRN, nLB and LLGC are more suitable for working with homogeneous data. However, transforming the heterogeneous information network into homogeneous subnetworks inevitably results in information loss. For example, in the *author-author* subnetwork, the venues where each author often publishes papers, and the terms that each author likes to use, are no longer known. Overall, GNetMine performs the second best by explicitly respecting the type differences in links and objects and thus encoding the typed information in the heterogeneous network in an organized way. Compared with GNetMine, RankClass achieves 16.6%, 0.58%, and 17.3% relative error reduction on the average classification accuracy when classifying authors, papers and venues, respectively. Although RankClass has a knowledge propagation framework similar to that of GNetMine, RankClass aims to compute the within-class ranking distribution to characterize each class, and it further employs the ranking results to iteratively extract the subnetwork corresponding to each specific class, and therefore, leads to more accurate knowledge propagation for each class.

3.4.3 CASE STUDY

In this section, we present a simple case study by listing the top-ranked data objects within each class. Recall that GNetMine performs the second best in the classification accuracy, and can generate a confidence score for each object related to each class. Thus, we can also rank data objects according to the confidence scores related to each class as the within-class ranking. In Tables 3.6 and 3.7, we show the comparison of the ranking lists of venues and terms generated by RankClass and GNetMine, respectively, with (0.5%, 0.5%) authors and papers labeled.

From comparing the ranking lists of the two types of objects, we can see that RankClass generates more meaningful ranking results than GNetMine. There is a high degree of consensus between the ranking list of venues generated by RankClass and the top venues in each research area. Similarly, the highly ranked terms generated by RankClass are in high agreement with the most representative keywords in each field. The reason why GNetMine fails to generate meaningful ranking lists is that the portions of labeled authors and papers are too limited to capture the distribution of the confidence score with regard to each class. In contrast, RankClass boosts the ranking performance by iteratively obtaining the clean subnetwork corresponding to each class, which favors the within-class ranking.

PART II

Meta-Path-Based Similarity Search and Mining

Meta-Path-Based Similarity Search

We now introduce a systematic approach for dealing with general heterogeneous information networks with a specified but arbitrary network schema, using a meta-path-based methodology. Under this framework, similarity search (Chapter 4) and other mining tasks such as relationship prediction (Chapter 5) can be addressed by systematic exploration of the network meta structure.

4.1 OVERVIEW

Similarity search, which aims at locating the most relevant information for a query in a large collection of datasets, has been widely studied in many applications. For example, in spatial database, people are interested in finding the k nearest neighbors for a given spatial object [35]; in information retrieval, it is useful to find similar documents for a given document or a given list of keywords. Object similarity is also one of the most primitive concepts for object clustering and many other data mining functions.

In a similar context, it is critical to provide effective similarity search functions in information networks, to find similar entities for a given entity. In a bibliographic network, a user may be interested in the top-k most similar authors for a given author, or the most similar venues for a given venue. In a network of tagged images such as Flickr, a user may be interested in search for the most similar pictures for a given picture. In an e-commerce system, a user would be interested in search for the most similar products for a given product. Different from the attribute-based similarity search, links play an essential role for similarity search in information networks, especially when the full information about attributes for objects is difficult to obtain.

There are a few studies leveraging link information in networks for similarity search, but most of these studies are focused on homogeneous networks or bipartite networks, such as personalized PageRank (P-PageRank) [29] and SimRank [28]. These similarity measures disregard the subtlety of different types among objects and links. Adoption of such measures to heterogeneous networks has significant drawbacks: even if we just want to compare objects of the same type, going through link paths of different types leads to rather different semantic meanings, and it makes little sense to mix them up and measure the similarity without distinguishing their semantics. For example, Table 4.1 shows the top-4 most similar venues for a given venue, DASFAA, based on (a) the common authors shared by two venues, or (b) the common topics (i.e., terms) shared by two venues. These two scenarios are represented by two distinct meta-paths: (a) $V P A P V$, denoting that the similarity is defined by the connection path "venue-paper-author-paper-venue," whereas (b) $V P T P V$, by the connection path "venue-paper-topic-paper-venue." A user can choose either (a) or (b) or their combination based on the preferred similarity semantics. According to Path (a), DASFAA is closer to DEXA, WAIM, and APWeb, that is, those that share many common authors, whereas according to Path (b), it is closer to Data Knowl. Eng., ACM Trans. DB Syst., and Inf. Syst., that is, those that address many common topics. Obviously, different connection paths lead to different semantics of similarity definitions, and produce rather different ranking lists even for the same query object.

Table 4.1: Top-4 most similar venues to "DASFAA" with two meta-paths

Rank	path: $V P A P V$	path: $V P T P V$
1	DASFAA	DASFAA
2	DEXA	Data Knowl. Eng.
3	WAIM	ACM Trans. DB Syst.
4	APWeb	Inf. Syst.

To systematically distinguish the semantics among paths connecting two objects, we introduce a meta-path-based similarity framework for objects of the same type in a heterogeneous network. A meta-path is a sequence of relations between object types, which defines a new composite relation between its starting type and ending type. The meta-path framework provides a powerful mechanism for a user to select an appropriate similarity semantics, by choosing a proper meta-path, or learn it from a set of training examples of similar objects.

In this chapter, we introduce the meta-path-based similarity framework, and relate it to two well-known existing link-based similarity functions for homogeneous information networks. Especially, we define a novel similarity measure, PathSim, that is able to find peer objects that are not only strongly connected with each other but also share similar visibility in the network. Moreover, we propose an efficient algorithm to support online top-k queries for such similarity search.

4.2 PATHSIM: A META-PATH-BASED SIMILARITY MEASURE

The similarity between two objects in a link-based similarity function is determined by how the objects are connected in a network, which can be described using paths. For example, in a co-author network, two authors can be connected either directly or via common co-authors, which are length-1 and length-2 paths, respectively. In a heterogeneous information network, however, due to the heterogeneity of the types of links, the way to connect two objects can be much more diverse. For example, in Table 4.2, Column I gives several path instances between authors in a bibliographic network, indicating whether the two authors have co-written a paper, whereas Column II gives several path instances between authors following a different connection path, indicating whether the two authors have ever published papers in the same venue. These two types of connections represent different relationships between authors, each having some different semantic meaning.

Table 4.2: Path instance vs. meta-path in heterogeneous information networks

	Column I: Connection Type I	Column II: Connection Type II
Path instance	Jim-P_1-Ann Mike-P_2-Ann Mike-P_3-Bob	Jim-P_1-SIGMOD-P_2-Ann Mike-P_3-SIGMOD-P_2-Ann Mike-P_4-KDD-P_5-Bob
Meta-path	Author-Paper-Author	Author-Paper-Venue-Paper-Author

Now the questions is, given an arbitrary heterogeneous information network, is there any way to systematically identify all the possible connection types (i.e., relations) between two object types? In order to do so, we propose two important concepts in the following.

4.2.1 NETWORK SCHEMA AND META-PATH

First, given a complex heterogeneous information network, it is necessary to provide its meta level (i.e., schema-level) description for better understanding the network. Therefore, we propose the concept of **network schema** to describe the meta structure of a network. The formal definition of network schema has been given in Definition 1.2 in Chapter 1.

The concept of network schema is similar to that of the ER (Entity-Relationship) model in database systems, but only captures the entity type and their binary relations, without considering the attributes for each entity type. Network schema serves as a template for a network, and tells how many types of objects there are in the network and where the possible links exist. Note that although a relational database can often be transformed into an information network, the latter is more general and can handle more unstructured and non-normalized data and links, and is also easier to deal with graph operations such as calculating the number of paths between two objects.

As we illustrated previously, two objects can be connected via different paths in a heterogeneous information network. For example, two authors can be connected via "author-paper-author" path, "author-paper-venue-paper-author" path, and so on. Formally, these paths are called *meta-paths*, defined as follows.

Definition 4.1 (Meta-path) A *meta-path* $\mathcal{P}$ is a path defined on the graph of network schema $T_G = (\mathcal{A}, \mathcal{R})$, and is denoted in the form of $A_1 \xrightarrow{R_1} A_2 \xrightarrow{R_2} \ldots \xrightarrow{R_l} A_{l+1}$, which defines a composite relation $R = R_1 \circ R_2 \circ \ldots \circ R_l$ between types A_1 and A_{l+1}, where $\circ$ denotes the composition operator on relations.

For the bibliographic network schema shown in Figure 4.1 (a), we list two examples of meta-paths in Figure 4.1 (b) and (c), where an arrow explicitly shows the direction of a relation. We say a path $p = (a_1 a_2 \ldots a_{l+1})$ between a_1 and a_{l+1} in network G follows the meta-path $\mathcal{P}$, if $\forall i$, $a_i \in A_i$ and each link $e_i = \langle a_i a_{i+1} \rangle$ belongs to each relation R_i in $\mathcal{P}$. We call these paths as *path instances* of $\mathcal{P}$, denoted as $p \in \mathcal{P}$. The examples of path instances have been shown in Table 4.2.

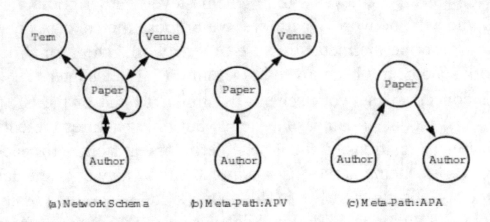

(a) Network Schema   (b) Meta-Path: APV   (c) Meta-Path: APA

Figure 4.1: Bibliographic network schema and meta-paths.

In addition to pointing out the meta-path we are interested in, we also need to consider how to quantify the connection between two objects following a given meta-path. Analogously, a meta-path-based measure in an information network corresponds to a feature in a traditional data set, which can be used in many mining tasks.

4.2.2 META-PATH-BASED SIMILARITY FRAMEWORK

Given a user-specified meta-path, say $\mathcal{P} = (A_1 A_2 \ldots A_l)$, several similarity measures can be defined for a pair of objects $x \in A_1$ and $y \in A_l$, according to the path instances between them following the meta-path. We use $s(x, y)$ to denote the similarity between x and y, and list several straightforward measures in the following.

- Path count: the number of path instances p between x and y following $\mathcal{P}$: $s(x, y) = |\{p : p \in \mathcal{P}\}|$.

- Random walk: $s(x, y)$ is the probability of the random walk that starts form x and ends with y following meta-path $\mathcal{P}$, which is the sum of the probabilities of all the path instances $p \in \mathcal{P}$ starting with x and ending with y, denoted as $Prob(p)$: $s(x, y) = \sum_{p \in \mathcal{P}} Prob(p)$.

- Pairwise random walk: for a meta-path $\mathcal{P}$ that can be decomposed into two shorter meta-paths with the same length $\mathcal{P} = (\mathcal{P}_1 \mathcal{P}_2)$, $s(x, y)$ is then the pairwise random walk probability starting from objects x and y and reaching the same middle object: $s(x, y) = \sum_{(p_1 p_2) \in (\mathcal{P}_1 \mathcal{P}_2)} Prob(p_1) Prob(p_2^{-1})$, where $Prob(p_1)$ and $Prob(p_2^{-1})$ are random walk probabilities of the two path instances.

In general, we can define a meta-path-based similarity framework for two objects x and y as: $s(x, y) = \sum_{p \in \mathcal{P}} f(p)$, where $f(p)$ is a measure defined on the path instance p between x and y. Note that P-PageRank and SimRank, two well-known network similarity functions, are weighted combinations of random walk measure or pairwise random walk measure, respectively, over meta-paths with different lengths in homogeneous networks. In order to use P-PageRank and SimRank in heterogeneous information networks, we need to specify the meta-path(s) we are interested in and restrict the random walk on the given meta-path(s).

4.2.3 PATHSIM: A NOVEL SIMILARITY MEASURE

Although there have been several similarity measures as presented above, they are biased to either highly visible objects or highly concentrated objects but cannot capture the semantics of peer similarity. For example, the path count and random walk-based similarity always favor objects with large degrees, and the pairwise random walk-based similarity favors concentrated objects where the majority of the links goes to a small portion of objects. However, in many scenarios, finding similar objects in networks is to *find similar peers*, such as finding similar authors based on their fields and reputation, finding similar actors based on their movie styles and productivity, and finding similar products based on their functions and popularity.

This motivated us to propose a new, meta-path-based similarity measure, called *PathSim*, that captures the subtlety of peer similarity. The intuition behind it is that two similar peer objects should not only be strongly connected, but also share comparable visibility. As the relation of peer should be symmetric, we confine PathSim to symmetric meta-paths. It is easy to see that, *round trip meta-paths* in the form of $\mathcal{P} = (\mathcal{P}_l \mathcal{P}_l^{-1})$ are always symmetric.

Definition 4.2 (PathSim: A meta-path-based similarity measure) Given a symmetric meta-path $\mathcal{P}$, PathSim between two objects x and y of the same type is:

$$s(x, y) = \frac{2 \times |\{p_{x \rightsquigarrow y} : p_{x \rightsquigarrow y} \in \mathcal{P}\}|}{|\{p_{x \rightsquigarrow x} : p_{x \rightsquigarrow x} \in \mathcal{P}\}| + |\{p_{y \rightsquigarrow y} : p_{y \rightsquigarrow y} \in \mathcal{P}\}|},$$

where $p_{x \rightsquigarrow y}$ is a path instance between x and y, $p_{x \rightsquigarrow x}$ is that between x and x, and $p_{y \rightsquigarrow y}$ is that between y and y.

This definition shows that given a meta-path $\mathcal{P}$, $s(x, y)$ is defined in terms of two parts: (1) their connectivity defined by the number of paths between them following $\mathcal{P}$; and (2) the balance of their visibility, where the visibility of an object according $\mathcal{P}$ is defined as the number of path instances between the object itself following $\mathcal{P}$. Note that we do count multiple occurrences of a path instance as the weight of the path instance, which is the product of weights of all the links in the path instance.

Table 4.3 presents in three measures the results of finding top-5 similar authors for "Anhai Doan," who is an established young researcher in the database field, under the meta-path $A\ P\ V\ P\ A$ (based on their shared venues), in the database and information system (DBIS) area. P-PageRank returns the most similar authors as those published substantially in the area, that is, highly ranked authors; SimRank returns a set of authors that are concentrated on a small number of venues shared with Doan; whereas PathSim returns Patel, Deshpande, Yang and Miller, who share very similar publication records and are also rising stars in the database field as Doan. Obviously, PathSim captures desired semantic similarity as peers in such networks.

Table 4.3: Top-5 similar authors for "AnHai Doan" in the $D\ B\ I\ S$ area

Rank	P-PageRank	SimRank	PathSim
1	AnHai Doan	AnHai Doan	AnHai Doan
2	Philip S. Yu	Douglas W. Cornell	Jignesh M. Patel
3	Jiawei Han	Adam Silberstein	Amol Deshpande
4	Hector Garcia-Molina	Samuel DeFazio	Jun Yang
5	Gerhard Weikum	Curt Ellmann	Renée J. Miller

The calculation of PathSim between any two objects of the same type given a certain meta-path involves matrix multiplication. Given a network $G = (\mathcal{V}, \mathcal{E})$ and its network schema T_G, we call the new adjacency matrix for a meta-path $\mathcal{P} = (A_1 A_2 \ldots A_l)$ a *relation matrix*, and is defined as $M = W_{A_1 A_2} W_{A_2 A_3} \cdots W_{A_{l-1} A_l}$, where $W_{A_i A_j}$ is the adjacency matrix between type A_i and type A_j. $M(i, j)$ represents the number of path instances between object $x_i \in A_1$ and object $y_j \in A_l$ under meta-path $\mathcal{P}$.

For example, relation matrix M for the meta-path $\mathcal{P} = (A\ P\ A)$ is a co-author matrix, with each element representing the number of co-authored papers for the pair of authors. Given a symmetric meta-path $\mathcal{P}$, PathSim between two objects x_i and x_j of the same type can be calculated as $s(x_i, x_j)$ $= \frac{2M_{ij}}{M_{ii}+M_{jj}}$, where M is the relation matrix for the meta-path $\mathcal{P}$, M_{ii} and M_{jj} are the visibility for x_i and x_j in the network given the meta-path.

It is easy to see that the relation matrix for the reverse meta-path of $\mathcal{P}_l$, which is $\mathcal{P}_l^{-1}$, is the *transpose* of relation matrix for $\mathcal{P}_l$. In this paper, we only consider the meta-path in the round trip form of $\mathcal{P} = (\mathcal{P}_l \mathcal{P}_l^{-1})$, to guarantee its symmetry and therefore the symmetry of the PathSim measure. By viewing PathSim in the meta-path-based similarity framework, $f(p) = 2\frac{w(a_1,a_2)...w(a_{l-1},a_l)}{M_{ii}+M_{jj}}$, for any path instance p starting from x_i and ending with x_j following the meta-path ($a_1 = x_i$ and $a_l = x_j$), where $w(a_m, a_n)$ is the weight for the link $\langle a_m, a_n \rangle$ defined in the adjacency matrix.

Some good properties of PathSim, such as symmetric, self-maximum, and balance of visibility, are shown in Theorem 4.3. For the balance property, we can see that the larger the difference of the visibility of the two objects, the smaller the upper bound for their PathSim similarity.

Theorem 4.3 (Properties of PathSim)

1. **Symmetric.** $s(x_i, x_j) = s(x_j, x_i)$.

2. **Self-maximum.** $s(x_i, x_j) \in [0, 1]$, and $s(x_i, x_i) = 1$.

3. **Balance of visibility.** $s(x_i, x_j) \leq \frac{2}{\sqrt{M_{ii}/M_{jj}}+\sqrt{M_{jj}/M_{ii}}}$.

Although using meta-path-based similarity we can define similarity between two objects given *any* round trip meta-paths, the following theorem tells us a *very long* meta-path is not very meaningful. Indeed, due to the sparsity of real networks, objects that are similar may share no immediate neighbors, and longer meta-paths will propagate similarities to remote neighborhoods. For example, as in the DBLP example, if we consider the meta-path *A P A*, only two authors that are co-authors have a non-zero similarity score; but if we consider longer meta-paths like *A P V P A* or *A P T P A*, authors will be considered to be similar if they have published papers in a similar set of venues or sharing a similar set of terms no matter whether they have co-authored. But how far should we keep going? The following theorem tells us that a very long meta-path may be misleading. We now use $\mathcal{P}^k$ to denote a meta-path repeating k times of the basic meta-path pattern of $\mathcal{P}$, e.g., $(A\,V\,A)^2 = (A\,V\,A\,V\,A)$.

Theorem 4.4 (Limiting behavior of PathSim under infinity-length meta-path) Let meta-path $\mathcal{P}^{(k)} = (\mathcal{P}_l \mathcal{P}_l^{-1})^k$, $M_{\mathcal{P}}$ be the relation matrix for meta-path $\mathcal{P}_l$, and $M^{(k)} = (M_{\mathcal{P}} M_{\mathcal{P}}^T)^k$ be the relation matrix for $\mathcal{P}^{(k)}$, then by PathSim, the similarity between objects x_i and x_j as $k \to \infty$ is:

$$\lim_{k\to\infty} s^{(k)}(i,j) = \frac{2\mathbf{r}(i)\mathbf{r}(j)}{\mathbf{r}(i)\mathbf{r}(i) + \mathbf{r}(j)\mathbf{r}(j)} = \frac{2}{\frac{\mathbf{r}(i)}{\mathbf{r}(j)} + \frac{\mathbf{r}(j)}{\mathbf{r}(i)}},$$

where $\mathbf{r}$ is the primary eigenvector of M, and $\mathbf{r}(i)$ is the i_{th} item of $\mathbf{r}$.

As primary eigenvectors can be used as authority ranking of objects [66], the similarity between two objects under an infinite meta-path can be viewed as a measure defined on their rankings ($\mathbf{r}(i)$ is the ranking score for object x_i). Two objects with more similar ranking scores will have higher similarity (e.g., SIGMOD will be similar to AAAI). Later experiments (Table 4.9) will show that this similarity, with the meaning of global ranking, is not that useful. Note that, the convergence of PathSim with respect to path length is usually very fast and the length of 10 for networks of the scale of DBLP can almost achieve the effect of a meta-path with an infinite length. Therefore, in this paper, we only aim at solving the top-k similarity search problem for a *relatively short* meta-path.

Even for a relatively short length, it may still be inefficient in both time and space to materialize all the meta-paths. Thus, we propose in Section 4.3 materializing relation matrices for short length meta-paths, and concatenating them online to get longer ones for a given query.

4.3 ONLINE QUERY PROCESSING FOR SINGLE META-PATH

Compared with P-PageRank and SimRank, the calculation for PathSim is much more efficient, as it is a local graph measure. But it still involves expensive matrix multiplication operations for top-k search functions, as we need to calculate the similarity between a query and every object of the same type in the network. One possible solution is to materialize all the meta-paths within a given length. Unfortunately, it is time and space expensive to materialize all the possible meta-paths. For example, in the DBLP network, the similarity matrix corresponding to a length-4 meta-path, $A\,P\,V\,P\,A$, for identifying similar authors publishing in common venues is a $710K \times 710K$ matrix, whose non-empty elements reaches $5G$, and requires storage size more than $40GB$.

In order to support fast online query processing for large-scale networks, we propose a methodology that partially materializes short length meta-paths and then concatenates them online to derive longer meta-path-based similarity. First, a baseline method (*PathSim-baseline*) is proposed, which computes the similarity between query object x and all the candidate objects y of the same type. Next, a co-clustering based pruning method (*PathSim-pruning*) is proposed, which prunes candidate objects that are not promising according to their similarity upper bounds. Both algorithms return *exact* top-k results for the given query. Note that the same methodology can be adopted by other meta-path-based similarity measures, such as random walk and pairwise random walk, by taking a different definition of similarity matrix accordingly.

4.3.1 SINGLE META-PATH CONCATENATION

Given a meta-path $\mathcal{P} = (\mathcal{P}_l \mathcal{P}_l^{-1})$, where $\mathcal{P}_l = (A_1 \cdots A_l)$, the relation matrix for path $\mathcal{P}_l$ is $M_{\mathcal{P}} = W_{A_1 A_2} W_{A_2 A_3} \cdots W_{A_{l-1} A_l}$, then the relation matrix for path $\mathcal{P}$ is $M = M_{\mathcal{P}} M_{\mathcal{P}}^T$. Let n be the number of objects in A_1. For a query object $x_i \in A_1$, if we compute the top-k most similar objects $x_j \in A_1$ for x_i on-the-fly, without materializing any intermediate results, computing M from scratch would be very expensive. On the other hand, if we have pre-computed and stored the relation matrix $M = M_{\mathcal{P}} M_{\mathcal{P}}^T$, it would be a trivial problem to get the query results. We only need to locate the corresponding row in the matrix for the query x_i, re-scale it using $(M_{ii} + M_{jj})/2$, and finally sort the new vector and return the top-k objects. However, fully materializing the relation matrices for all possible meta-paths is also impractical, since the space complexity $(O(n^2))$ would prevent us from storing M for every meta-path. Instead of taking the above extremes, we partially materialize relation matrix $M_{\mathcal{P}}^T$ for meta-path $\mathcal{P}_l^{-1}$, and compute top-k results online by concatenating $\mathcal{P}_l$ and $\mathcal{P}_l^{-1}$ into $\mathcal{P}$ without full matrix multiplication.

We now examine the concatenation problem, that is, when the relation matrix M for the full meta-path $\mathcal{P}$ is not pre-computed and stored, but the relation matrix $M_{\mathcal{P}}^T$ corresponding to the partial meta-path $\mathcal{P}_l^{-1}$ is available. In this case, we assume the main diagonal of M, that is, $D = (M_{11}, \ldots, M_{nn})$, is pre-computed and stored. Since for $M_{ii} = M_{\mathcal{P}}(i, :)M_{\mathcal{P}}(i, :)^T$, the calculation only involves $M_{\mathcal{P}}(i, :)$ itself, and only $O(nd)$ in time and $O(n)$ in space are required, where d is the average number of non-zero elements in each row of $M_{\mathcal{P}}$ for each object.

In this study, we only consider concatenating the partial paths $\mathcal{P}_l$ and $\mathcal{P}_l^{-1}$ into the form $\mathcal{P} = \mathcal{P}_l\mathcal{P}_l^{-1}$ or $\mathcal{P} = \mathcal{P}_l^{-1}$. For example, given a pre-stored meta-path $A\,P\,V$, we are able to answer queries for meta-paths $A\,P\,V\,P\,A$ and $VPAPV$. For our DBLP network, to store relation matrix for partial meta-path $A\,P\,V$ only needs around $25M$ space, which is less than 0.1% of the space for materializing meta-path $A\,P\,V\,P\,A$. Other concatenation forms that may lead to different optimization methods are also possible (e.g., concatenating several short meta-paths). In the following discussion, we focus on the algorithms using the concatenation form $\mathcal{P} = \mathcal{P}_l\mathcal{P}_l^{-1}$.

4.3.2 BASELINE

Suppose we know the relation matrix $M_{\mathcal{P}}$ for meta-path P_l, and the diagonal vector $D = (M_{ii})_{i=1}^n$, in order to get top-k objects $x_j \in A_1$ with the highest similarity for the query x_i, we need to compute $s(i, j)$ for all x_j. The straightforward baseline is: (1) first apply vector-matrix multiplication to get $M(i, :) = M_{\mathcal{P}}(i, :)M_{\mathcal{P}}^T$; (2) calculate $s(i, j) = \frac{2M(i,j)}{M(i,i)+M(j,j)}$ for all $x_j \in$

A_1; and (3) sort $s(i, j)$ to return the top-k list in the final step. When n is very large, the vector-matrix computation will be too time consuming to check every possible object x_j. Therefore, we first select x_j's that are not orthogonal to x_i in the vector form, by following the links from x_i to find 2-step neighbors in relation matrix $M_\mathcal{P}$, that is, $x_j \in CandidateSet = \{\bigcup_{y_k \in M_\mathcal{P}.neighbors(x_i)} M_\mathcal{P}^T.neighbors(y_k)\}$, where $M_\mathcal{P}.neighbors(x_i) = \{y_k | M_\mathcal{P}(x_i, y_k) \neq 0\}$, which can be easily obtained in the sparse matrix form of $M_\mathcal{P}$ that indexes both rows and columns. This will be much more efficient than pairwise comparison between the query and all the objects of that type. We call this baseline concatenation algorithm as *PathSim-baseline*.

The *PathSim-baseline* algorithm, however, is still time consuming if the candidate set is large. Although $M_\mathcal{P}$ can be relatively sparse given a short length meta-path, after concatenation, M could be dense, i.e., the *CandidateSet* could be very large. Still, considering the query object and one candidate object represented by query vector and candidate vector, the dot product between them is proportional to the size of their non-zero elements. The time complexity for computing PathSim for each candidate is $O(d)$ on average and $O(m)$ in the worst case, that is, $O(nm)$ in the worst case for all the candidates, where n is the row size of $M_\mathcal{P}$ (i.e., the number of objects in type A_1), m the column size of $M_\mathcal{P}$ (i.e., the number of objects in type A_l), and d the average non-zero element for each object in $M_\mathcal{P}$. We now propose a co-clustering based top-k concatenation algorithm, by which non-promising target objects are dynamically filtered out to reduce the search space.

4.3.3 CO-CLUSTERING-BASED PRUNING

In the baseline algorithm, the computational costs involve two factors. First, the more candidates to check, the more time the algorithm will take; second, for each candidate, the dot product of query vector and candidate vector will at most involve m operations, where m is the vector length. The intuition to speed up the search is to prune unpromising candidate objects using simpler calculations. Based on the intuition, we propose a co-clustering-based (i.e., clustering rows and columns of a matrix simultaneously) path concatenation method, which first generates co-clusters of two types of objects for partial relation matrix, then stores necessary statistics for each of the blocks corresponding to different co-cluster pairs, and then uses the block statistics to prune the search space. For better illustration, we call clusters of type A_1 as **target clusters**, since the objects in A_1 are the targets for the query; and call clusters of type A_l as **feature clusters**, since the objects in A_l serve as features to calculate the similarity between the query and the target objects. By partitioning A_1 into different target clusters, if a whole target cluster is not similar to the query, then all the objects in the target cluster are likely not in the final top-k lists and can be pruned. By partitioning A_l into different feature clusters, cheaper calculations on the dimension-reduced query vector and candidate vectors can be used to derive the similarity upper bounds. This pruning idea is illustrated in Figure 4.2 using a toy example with 9 target objects and 6 feature objects. The readers may refer to the PathSim paper [65] for the concrete formulas of the upper bounds and their derivations.

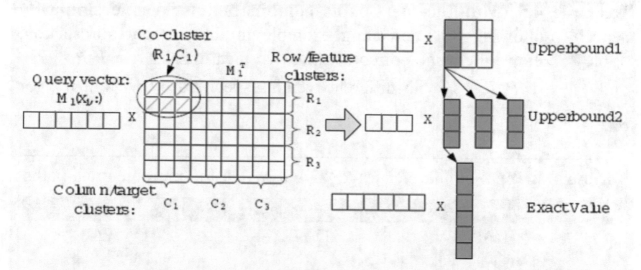

Figure 4.2: Illustration of pruning strategy. Given the partial relation matrix M_l^T and its 3×3 co-clusters, and the query vector $M_l(x_i, :)$ for query object x_i, first the query vector is compressed into the aggregated query vector with the length of 3, and the upper bounds of the similarity between the query and all the 3 target clusters are calculated based on the aggregated query vector and aggregated cluster vectors; second, for each of the target clusters, if they cannot be pruned, calculate the upper bound of the similarity between the query and each of the 3 candidates within the

cluster using aggregated vectors; third, if the candidates cannot be pruned, calculate the exact similarity value using the non-aggregated query vector and candidate vectors.

Experiments show that *PathSim-Pruning* can significantly improve the query processing speed comparing with the baseline algorithm, without affecting the search quality.

4.4 MULTIPLE META-PATHS COMBINATION

In Section 4.3, we presented algorithms for similarity search using single meta-path. Now, we present a solution to combine multiple meta-paths. Formally, given r round trip meta-paths from Type A back to Type A, $\mathcal{P}_1$, $\mathcal{P}_2, \ldots, \mathcal{P}_r$, and their corresponding relation matrix $M_1, M_2, \ldots, M_r$, with weights $w_1, w_2, \ldots, w_r$ specified by users, the combined similarity between objects x_i, $x_j \in A$ are defined as: $s(x_i, x_j) = \sum_{l=1}^{r} w_l s_l(x_i, x_j)$, where $s_l(x_i, x_j) = \frac{2M_l(i,j)}{M_l(i,i) + M_l(j,j)}$.

Example 4.5 (Multiple meta-paths combination for venue similarity search) Following the motivating example in the introduction section, Table 4.4 shows the results of combining two meta-paths $\mathcal{P}_1 = V P A P V$ and $\mathcal{P}_2 = V P T P V$ with different weights specified by w_1 and w_2, for query "DASFAA."

Table 4.4: Top-5 similar venues to "DASFAA" using multiple meta-paths

Rank	$w_1 = 0.2, w_2 = 0.8$	$w_1 = 0.5, w_2 = 0.5$	$w_1 = 0.8, w_2 = 0.2$
1	DASFAA	DASFAA	DASFAA
2	Data Knowl. Eng.	DEXA	DEXA
3	CIKM	CIKM	WAIM
4	EDBT	Data Knowl. Eng.	CIKM
5	Inf. Syst.	EDBT	APWeb

The reason why we need to combine several meta-paths is that each meta-path provides a unique angle (or a unique feature space) to view the similarity between objects, and the ground truth may be a cause of different factors. Some useful guidance of the weight assignment includes: longer meta-path utilize more remote relationships and thus should be assigned with a smaller weight, such as in P-PageRank and SimRank; and, meta-paths with more important relationships should be assigned with a higher weight. For automatically determining the weights, users could provide training examples of similar objects to learn the weights of different meta-paths using learning algorithms.

We now evaluate the quality of similarity measure generated by combined meta-paths, according to their performance for clustering tasks in the "*four-area*" dataset. First, two meta-paths for the venue type, namely, $V A V$ and $V T V$ (short for $V P A P V$ and $V P T P V$), are selected and their linear combinations with different weights are considered. Second, two meta-paths with the same basic path but different lengths, namely $A V A$ and $(A V A)^2$, are selected and their linear combinations with different weights are considered. The clustering accuracy measured by NMI for conferences and authors is shown in Table 4.5, which shows that the combination of multiple meta-paths can produce better similarity than the single meta-path in terms of clustering accuracy.

Table 4.5: Clustering accuracy for PathSim for meta-path combinations on the "*four-area*"

w_1	0	0.2	0.4	0.6	0.8	1
w_2	1	0.8	0.6	0.4	0.2	0
$V A V$; $V T V$	0.7917	0.7936	0.8299	0.8587	0.8123	0.8116
$A V A$; $(A V A)^2$	0.6091	0.6219	0.6506	0.6561	0.6508	0.6501

4.5 EXPERIMENTAL RESULTS

To show the effectiveness of the PathSim measure and the efficiency of the proposed algorithms, we use the bibliographic networks extracted from DBLP and Flickr in the experiments.

We use the DBLP dataset downloaded in November 2009 as the main test dataset. It contains over $710K$ authors, $1.2M$ papers, and $5K$ venues (conferences/journals). After removing stopwords in paper titles, we get around $70K$ terms appearing more than once. This dataset is referred as the *full-DBLP* dataset. Two small subsets of the data (to alleviate the high computational costs of P-PageRank and SimRank) are used for the comparison with other similarity measures in effectiveness: (1) the *DBIS* dataset, which contains all the 464 venues and top-5000 authors from the database and information system area; and (2) the *four-area* dataset, which contains 20 venues and top-5000 authors from 4 areas: *database, data mining, information retrieval*, and *artificial intelligence* [64], and cluster labels are given for all the 20 venues and a subset of 1713 authors.

Table 4.6: Case study of five similarity measures on query "PKDD" on the *DBIS* dataset

Rank	P-PageRank	SimRank	RW	PRW	PathSim
1	PKDD	PKDD	PKDD	PKDD	PKDD
2	KDD	Local Pattern Detection	KDD	Local Pattern Detection	ICDM
3	ICDE	KDID	ICDM	DB Support for DM Appl.	SDM
4	VLDB	KDD	PAKDD	Constr. Min. & Induc. DB	PAKDD
5	SIGMOD	Large-Scale Paral. DM	SDM	KDID	KDD
6	ICDM	SDM	TKDE	MCD	DMKD
7	TKDE	ICDM	SIGKDD Expl.	Pattern Detection & Disc.	SIGKDD Expl.
8	PAKDD	SIGKDD Expl.	ICDE	RSKD	Knowl. Inf. Syst.
9	SIGIR	Constr. Min. & Induc. DB	SEBD	WImBI	JIIS
10	CIKM	TKDD	CIKM	Large-Scale Paral. DM	KDID

For additional case studies, we construct a Flickr network from a subset of the Flickr data, which contains four types of objects: images, users, tags, and groups. Links exist between images and users, images and tags, and images and groups. We use 10,000 images from 20 groups as well as their related 664 users and 10,284 tags appearing more than once to construct the network.

4.5.1 EFFECTIVENESS

Comparing PathSim with other measures When a meta-path $\mathcal{P} = (\mathcal{P}_l \mathcal{P}_l^{-1})$ is given, other measures such as random walk (RW) and pairwise random walk (PRW) can be applied to the same meta-path, and P-PageRank and SimRank can be applied to the sub-network extracted from $\mathcal{P}$. For example, for the meta-path $V\ P\ A\ P\ V\ (V\ A\ V$ in short) for finding venues sharing the same set of authors, the bipartite graph M_{CA}, derived from the relation matrix corresponding to VPA can be used in both P-PageRank and SimRank algorithms. In our experiments, the damping factor for P-PageRank is set as 0.9 and that for SimRank is 0.8.

First, a case study is shown in Table 4.6, which is applied to the $D\ B\ I$ S dataset, under the meta-path $V\ A\ V$. One can see that for query "PKDD" (short for "Principles and Practice of Knowledge Discovery in Databases," a European data mining conference), P-PageRank favors the venues with higher visibility, such as KDD and several well-known venues; SimRank prefers more concentrated venues (i.e., a large portion of publications goes to a small set of authors) and returns many not well-known venues such as "Local Pattern Detection" and KDID; RW also favors highly visible objects such as KDD, but brings in fewer irrelevant venues due to that it utilizes merely one short meta-path; PRW performs similar to SimRank, but brings in more not so well-known venues due to the short meta-path it uses; whereas PathSim returns the venues in both the area and the reputation similar to PKDD, such as ICDM and SDM.

Table 4.7: Comparing the accuracy of top-15 query results for five similarity measures on the $D\ B\ I\ S$ dataset measured by nDCG

	P-PageRank	SimRank	RW	PRW	PathSim
Accuracy	0.5552	0.6289	0.7061	0.5284	0.7446

We then labeled top-15 results for 15 queries from the venues in the D $B\ I\ S$ dataset (i.e., SIGMOD, VLDB, ICDE, PODS, EDBT, DASFAA, KDD, ICDM, PKDD, SDM, PAKDD, WWW, SIGIR, TREC, and APWeb), to test the quality of the ranking lists given by 5 measures. We label each result object with a relevance score at one of the three levels: 0 (non-relevant), 1 (somewhat relevant), and 2 (very relevant). Then we use the measure nDCG (i.e., Normalized Discounted Cumulative Gain, with the value between 0 and 1, the higher the better) [27] to evaluate the quality of a ranking algorithm by comparing its output ranking results with the labeled ones (Table 4.7). The results show that PathSim gives the best ranking quality in terms of human intuition, which is consistent with the previous case study.

Table 4.8: Top-10 similar authors to "Christos Faloutsos" under different meta-paths on the *full-DBLP* dataset

Rank	Author	Rank	Author
1	Christos Faloutsos	1	Christos Faloutsos
2	Spiros Papadimitriou	2	Jiawei Han
3	Jimeng Sun	3	Rakesh Agrawal
4	Jia-Yu Pan	4	Jian Pei
5	Agma J. M. Traina	5	Charu C. Aggarwal
6	Jure Leskovec	6	H. V. Jagadish
7	Caetano Traina Jr.	7	Raghu Ramakrishnan
8	Hanghang Tong	8	Nick Koudas
9	Deepayan Chakrabarti	9	Surajit Chaudhuri
10	Flip Korn	10	Divesh Srivastava

(a) Path: APA (b) Path: $APVPA$

Semantic meanings of different meta-paths As we pointed out, different meta-paths give different semantic meanings, which is one of the reasons that similarity definitions in homogeneous networks cannot be applied directly to heterogeneous networks. Besides the motivating example in the introduction section, Table 4.8 shows the author similarity under two scenarios for author Christos Faloutsos: *co-authoring papers* and *publishing papers in the same venues*, represented by the meta-paths APA and $APVPA$, respectively. One can see that the first path returns co-authors who have strongest connections with Faloutsos (i.e., students and close collaborators) in DBLP, whereas $A\ P\ V\ P\ A$ returns those publishing papers in the most similar venues.

Table 4.9: Top-10 similar venues to "SIGMOD" under meta-paths with different lengths on the *full-DBLP* dataset

Rank	Venue	Score	Rank	Venue	Score	Rank	Venue	Score
1	SIGMOD	1	1	SIGMOD	1	1	SIGMOD	1
2	VLDB	0.981	2	VLDB	0.997	2	AAAI	0.9999
3	ICDE	0.949	3	ICDE	0.996	3	ESA	0.9999
4	TKDE	0.650	4	TKDE	0.787	4	ITC	0.9999
5	SIGMOD Record	0.630	5	SIGMOD Record	0.686	5	STACS	0.9997
6	IEEE Data Eng. Bul.	0.530	6	PODS	0.586	6	PODC	0.9996
7	PODS	0.467	7	KDD	0.553	7	NIPS	0.9993
8	ACM Trans. DB Sys.	0.429	8	CIKM	0.540	8	Comput. Geom.	0.9992
9	EDBT	0.420	9	IEEE Data Eng. Bul.	0.532	9	ICC	0.9991
10	CIKM	0.410	10	J. Comp. Sys. Sci.	0.463	10	ICDE	0.9984

(a) Path: $(VPAPV)^2$ (b) Path: $(VPAPV)^4$ (c) Path: $(VPAPV)^\infty$

The impact of path length The next interesting question is how the length of meta-path impacts the similarity definition. Table 4.9 shows an example of venues similar to "SIGMOD" with three meta-paths, using exactly the same basic meta-path, but with different repeating times. These meta-paths are $(V\ P\ A\ P\ V)^2$, $(V\ P\ A\ P\ V)^4$, and its infinity form (global ranking-based similarity). Note that in $(V\ P\ A\ P\ V)^2$, two venues are similar if they share many similar authors who publish papers in *the same* venues; while in $(V\ P\ A\ P\ V)^4$, the similarity definition of those venues will be further relaxed, namely, two venues are similar if they share many similar authors who publish papers in *similar* venues. Since venue type only contains $5K$ venues, we are able to get the full materialization relation matrix for $(V\ P\ A\ P\ V)^2$. $(V\ P\ A\ P\ V)^4$ is obtained using meta-path concatenation from $(V\ P\ A\ P\ V)^2$. The results are summarized in Table 4.9, where longer paths gradually bring in more remote neighbors, with higher similarity scores, and finally, it degenerates into global ranking comparison. Through this study, one can see that a meta-path with relatively short length is good enough to measure similarity, where a long meta-path may even reduce the quality.

4.5.2 EFFICIENCY COMPARISON

The time complexity for SimRank is $O(K N^2 d^2)$, where K is the number of iterations, N is the total number of objects, and d is the average neighbor size; the time complexity for calculating P-PageRank for one query is $O(K N d)$, where K, N, d has the same meaning as in SimRank, whereas the time complexity for PathSim using *PathSim-baseline* for single query is $O(nd)$, where $n < N$ is the number of objects in the target type, d is the average degree of objects in target type for partial relation matrix $M_{\mathcal{P}_l}$. The time complexity for RW and PRW are the same as PathSim. We can see that similarity measure only using one meta-path is much more efficient than those also using longer meta-paths in the network (e.g., SimRank and P-PageRank).

Two algorithms, *PathSim-baseline* and *PathSim-pruning*, introduced in Section 4.3, are compared, for efficiency study under different meta-paths, namely, $V P A P V$ and $(V P A P V)^2$ (denoted as $V A V$ and $V A V A V$ for short). The results show that the denser the relation matrix corresponding to the partial meta-path ($M_{V P A P V}$ in comparison with $M_{V P A}$), the greater the pruning power. The improvement rates are 18.23% and 68.04% for the 2 meta-paths.

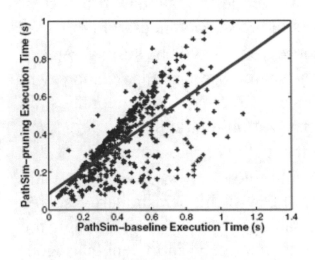

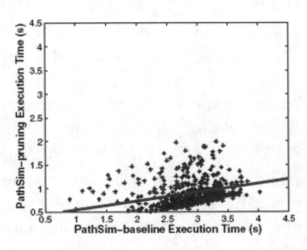

(a) Meta-path: $V A V$

(b) Meta-path: $V A V A V$

Figure 4.3: Pruning power denoted by the slope of the fitting line under two meta-paths for type conference on the *full-DBLP* dataset. Each dot represents a query under the indicated meta-path.

4.5.3 CASE-STUDY ON FLICKR NETWORK

In this case study, we show that to retrieve similar images for a query image one can explore links in the network rather than the content information. Let "I" represent images, "T" tags that associated with each image, and "G" groups that each image belongs to. Two meta-paths are used and compared. One is $I\,T\,I$, which means common tags are used by two images at evaluation of their similarity. The results are shown in Figure 4.4. The other is $I\,T\,I\,G\,I\,T\,I$, which means tags similarities are further measured by their shared groups, and two images can be similar even if they do not share many exact same tags as long as these tags are used by many images of the same groups. One can see that the second meta-path gives better results than the first, as shown in Figure 4.5, where the first image is the input query. This is likely due to that the latter meta-path provides additional information related to image groups, and thus improves the similarity measure between images.

(a) top-1   (b) top-2   (c) top-3   (d) top-4   (e) top-5   (f) top-6

Figure 4.4: Top-6 images in Flickr network under meta-path $I\,T\,I$.

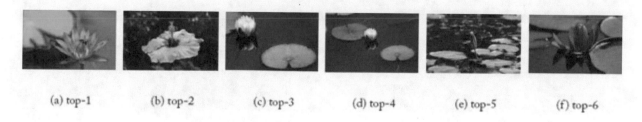

(a) top-1   (b) top-2   (c) top-3   (d) top-4   (e) top-5   (f) top-6

Figure 4.5: Top-6 images in Flickr network under meta-path $I\,T\,I\,G\,I\,T\,I$.

CHAPTER 5

Meta-Path-Based Relationship Prediction

In Chapter 4, we introduced a meta-path-based similarity measure, PathSim, for heterogeneous information networks. The concept of meta-path serves not only as a basis for similarity search but also as a key for mining and learning general heterogeneous networks with an arbitrary network schema, because this notion provides a way to guide us to systematically build link-based features. In this chapter, we examine a new mining task, relationship prediction in heterogeneous information networks, by exploring meta-path-based features.

5.1 OVERVIEW

Link prediction, that is, *predicting the emergence of links in a network based on certain current or historical network information,* has been a popular theme of research in recent years, thanks to the popularity of social networks and other online systems. The applications of link prediction range from social networks to biological networks, as it addresses the fundamental question of *whether* a link will form between two nodes in the future. Most of the existing link prediction methods [24; 38; 39; 40; 74] are designed for homogeneous networks, in which only one type of objects exists in the network. For example, in a friendship network or a co-author network, a user may like to predict possible new friendship between two persons or new co-authorship between two authors, based on the existing links in a network.

In the real world, most networks are heterogeneous, where multiple types of objects and links exist. In such networks, objects are connected by different types of relationships. Objects are connected together not only by immediate links, but also by more sophisticated relationships that follow some meta-path-based relations. Here we extend the *link prediction* problem in homogeneous information networks to the *relationship prediction* problem in heterogeneous information networks, where *a relationship could be an immediate link or a path instance following some meta-paths*. Many real-world problems can be considered as relationship prediction tasks, such as citation prediction in a bibliographic network, product recommendation in an e-commerce network, and online advertisement click prediction in an online system-based network.

The heterogeneity of objects and links makes it difficult to use well-known topological features in homogeneous networks for algorithmic design. For example, the number of the common neighbors is frequently used as a feature for link prediction in homogeneous networks. However, the neighbors of an object in a heterogeneous network often are of different types, and a simple measure like the number of shared neighbors cannot reflect this heterogeneity.

We thus propose a meta-path-based relationship prediction framework to overcome this difficulty. Instead of treating objects and links of different types equally or extracting homogeneous subnetworks from the original network, we propose a meta-path-based topological feature framework for heterogeneous networks. The goal is to systematically define the relations between objects encoded in different paths using the meta structure of these paths, that is, the meta-paths.

Two case studies using the meta-path-based relationship prediction framework are presented in this chapter. The first is on co-authorship prediction in the DBLP network, whereas the second proposes a novel prediction model that can predict when a relationship is going to built in a given heterogeneous information network.

5.2 META-PATH-BASED RELATIONSHIP PREDICTION FRAMEWORK

Different from traditional link prediction tasks for homogeneous information networks, in a heterogeneous information network scenario, it is necessary to specify which type of relationships to predict. The relationship to be predicted is called the *target relation* and can be described using a meta-path. For example, the relation *co-authorship* can be described as a meta-path $A - P - A$. Moreover, in order to build an effective prediction model, one need to examine how to construct the meta-path-based topological features between two objects for each potential relationship. In this section, we first examine how to systematically build topological feature space using meta-paths, and then present a supervised prediction framework where the meta-path-based topological measures are used as features.

5.2.1 META-PATH-BASED TOPOLOGICAL FEATURE SPACE

Topological features, also known as structural features, reflect the essential connectivity properties for pairs of objects. Topological feature-based link prediction aims at inferring the future connectivity by leveraging the current connectivity of the network. There are some frequently used topological features defined in homogeneous networks, such as the number of common neighbors, preferential attachment [5; 49], and $katz_\beta$ [33]. We first review several commonly used topological features in homogeneous networks, and then propose a systematic meta-path-based methodology to define topological features in heterogeneous networks.

Existing Topological Features

We introduce several well-known and frequently used topological features in homogeneous networks. For more topological features, the readers can refer to [39].

- *Common neighbors*. Common neighbors is defined as the number of common neighbors shared by two objects, a_i and a_j, namely $|\Gamma(a_i) \cap \Gamma(a_j)|$, where $\Gamma(a)$ is the notation for neighbor set of the object a and $|\cdot|$ denotes the size of a set.

- *Jaccard's coefficient*. Jaccard's coefficient is a measure to evaluate the similarity between two neighbor sets, which can be viewed as the normalized number of common neighbors, namely $\frac{|\Gamma(a_i) \cap \Gamma(a_j)|}{|\Gamma(a_i) \cup \Gamma(a_j)|}$.

- *Katz$_\beta$.* Katz$_\beta$ [33] is a weighted summation of counts of paths between two objects with different lengths, namely $\sum_{l=1}^{\infty} \beta^l |path_{a_i,a_j}^{\langle l \rangle}|$, where β^l is the damping factor for the path with length l.

- *PropFlow.* In a recent study [40], a random walk-based measure PropFlow is proposed to measure the topological feature between two objects. This method assigns the weighs to each path (with fixed length l) using the products of proportions of the flows on the edges.

One can see that most of the existing topological features in homogeneous networks are based on neighbor sets or paths between two objects. However, as there are multi-typed objects and multi-typed relations in heterogeneous networks, the neighbors of an object could belong to multiple types, and the paths between two objects could follow different meta-paths and indicate different relations. Thus, it is necessary to design a more complex strategy to generate topological features in heterogeneous networks.

Meta-Path-Based Topological Features

To design topological features in heterogeneous networks, we first define the topology between two objects using meta-paths, and then define measures on a specific topology. In other words, a meta-path-based topological feature space is comprised of two parts: the meta-path-based topology and the measure functions that quantify the topology.

Meta-path-based topology As introduced in Chapter 4, a meta-path is a path defined over a network schema and denotes a composition relation over a heterogeneous network. By checking the existing topological features defined in a homogeneous network, we can find that both the neighbor set-based features and path-based features can be generalized in the heterogeneous information network, by considering paths following different meta-paths. For example, if we treat each type of neighbors separately and extend the immediate neighbors to n-hop neighbors (i.e., the distance between one object and its neighbors are n), the common neighbor feature between two objects then becomes the count of paths between the two objects following different meta-paths. For path-based features, such as $Katz_\beta$, it can be extended as a combination of paths following different meta-paths, where each meta-path defines a unique topology between objects, representing a special relation.

Meta-paths between two object types can be obtained by traversing the graph of network schema, using standard traversal methods such as the BFS (breadth-first search) algorithm. As the network schema is a much smaller graph compared with the original network, this stage is very fast. We can enumerate all the meta-paths between two object types by setting a length constraint. For example, in order to predict co-authorship in the DBLP network, we extract all the meta-paths within a length constraint, say 4, starting and ending with the author type A. The meta-paths between authors up to length 4 are summarized in Table 5.1, where the semantic meaning of each relation denoted by each meta-path are given in the second column.

Table 5.1: Meta-paths under length 4 between authors in the DBLP network

Meta-path	Semantic Meaning of the Relation
$A - P - A$	a_i and a_j are co-authors
$A - P \rightarrow P - A$	a_i cites a_j
$A - P \leftarrow P - A$	a_i is cited by a_j
$A - P - V - P - A$	a_i and a_j publish in the same venues
$A - P - A - P - A$	a_i and a_j are co-authors of the same authors
$A - P - T - P - A$	a_i and a_j write the same topics
$A - P \rightarrow P \rightarrow P - A$	a_i cites papers that cite a_j
$A - P \leftarrow P \leftarrow P - A$	a_i is cited by papers that are cited by a_j
$A - P \rightarrow P \leftarrow P - A$	a_i and a_j cite the same papers

| $A - P \leftarrow P \rightarrow P - A$ | a_i and a_j are cited by the same papers |

Measure functions on meta-paths Once the topologies given by meta-paths are determined, the next stage is to propose measures to quantify these meta-paths for pairs of objects. Here we list four measures along the lines of topological features in homogeneous networks. They are *path count, normalized path count, random walk*, and *symmetric random walk*, defined as follows. Additional measures can be proposed, such as *pairwise random walk* mentioned in Chapter 4.

- **Path count**. Path count measures the number of path instances between two objects following a given meta-path R, denoted as $P C_R$. Path count can be calculated by the products of adjacency matrices associated with each relation in the meta-path.

- **Normalized path count**. Normalized path count is to discount the number of paths between two objects in the network by their overall connectivity, and is defined as $N P C_R(a_i, a_j) = \frac{PC_R(a_i,a_j)+PC_{R^{-1}}(a_j,a_i)}{Z_R(a_i,a_j)}$, where R^{-1} denotes the inverse relation of R, $Z_R(a_i, a_j)$ is some normalization factor. For example, PathSim [65] is a special case of normalized path count, where $Z_R(a_i, a_j) = P C_R(a_i, a_i) + P C_R(a_j, a_j)$ for symmetric R's.

- **Random walk**. Random walk measure along a meta-path is defined as $R W_R(a_i, a_j) = \frac{PC_R(a_i,a_j)}{PC_R(a_i,\cdot)}$, where $P C_R(a_i, \cdot)$ denotes the total number of paths following R starting with a_i, which is a natural generalization of PropFlow [40].

- **Symmetric random walk**. Symmetric random walk considers the random walk from two directions along the meta-path, and defined as $S R W_R(a_i, a_j) = R W_R(a_i, a_j) + R W_{R^{-1}}(a_j, a_i)$.

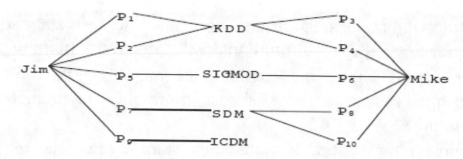

Figure 5.1: An example of path instances between two authors following *A-P-V-P-A*.

Taking the example in Figure 5.1, we show the calculation of these measures. Let R denote the relation represented by meta-path $A - P - V - P - A$. It is easy to check it is symmetric, i.e., $R = R^{-1}$. Let J denote Jim, and M denote Mike. We can see that $P C_R(J, M) = 7$, $N P C_R(J, M) = \frac{7+7}{7+9} = 7/8$ (under PathSim), $R W_R(J, M) = 1/2$, $R W_R(M, J) = 7/16$, and $S R W_R(J, M) = 15/16$.

For each meta-path, we can apply any measure functions on it and obtain a unique topological feature. So far, we have provided a systematic way to define the topological features in heterogeneous networks, which is a large space defined over *topology* × *measure*. These meta-path-based topological features can serve a good feature space for mining and learning tasks, such as relationship prediction.

5.2.2 SUPERVISED RELATIONSHIP PREDICTION FRAMEWORK

The supervised learning framework is summarized in Figure 5.2. Generally, given a past time interval $T_0 = [t_0, t_1)$, we want to use the topological features extracted from the aggregated network in the time period T_0, to predict the relationship building in a future time interval, say $T_1 = [t_1, t_2)$. In the **training stage**, we first sample a set of object pairs in T_0, collect their associated topological features represented as **x**'s in T_0, and record relationship building facts between them represented as y's in the future interval T_1. A training model is then built to learn the best coefficients associated with each topological feature by maximizing the likelihood of relationship building. In the **test stage**, we apply the learned coefficients to the topological features for the test pairs, and compare the predicted relationship with the ground truth. Note that the test stage may have different past interval T_0' and future interval T_1' as in the training stage, but we require they have the same lengths as the intervals in the training stage, namely using the same amount of past information to predict the same length of future.

For most of the existing link prediction studies, the tasks are predicting whether a new link will appear in the future. In other words, y is a binary variable and is usually modeled as following Bernoulli distribution. While in a more general case, y can be variables related to any reasonable value of the relationship for a pair of objects. For example, in order to predict when a relationship is going to be built, y could be modeled a positive real value following exponential distribution; in order to predict the frequency of a relationship (e.g., how many times two authors are going to collaborate), y could be modeled as a non-negative integer following Poisson distribution. Then statistical models can be built based on the distribution assumptions of y, such as logistic regression model for binary variables and generalized linear model for more sophisticated assumptions.

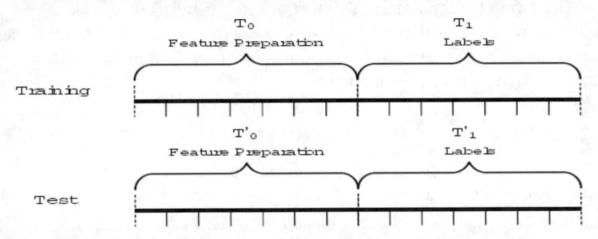

Figure 5.2: Supervised framework for relationship prediction.

Two case studies of relationship prediction are shown in the following sections, both of which follow the supervised relationship prediction framework, but with different purposes and thus different assumptions on the response variable y.

5.3 CO-AUTHORSHIP PREDICTION

For the first case study, we study the problem of co-authorship prediction in the DBLP bibliographic network, that is, whether two authors are going to collaborate in a future interval for the first time. In this case, the target relation for prediction is co-authorship relation, which can be described using meta-path $A - P - A$. For the topological features, we study all the meta-path listed in Table 5.1 other than $A - P - A$ and all the measures listed in the last section.

We next introduce the relationship prediction model which models the probability of co-authorship between two authors as a function of topological features between them. Given the training pairs of authors, we first extract the topological features for them, and then build the prediction model to learn the weights associated with these features.

5.3.1 THE CO-AUTHORSHIP PREDICTION MODEL

In order to predict whether two authors are going to collaborate in a future interval, denoted as y, we use the logistic regression model as the prediction model. For each training pair of authors $\langle a_{i_1}, a_{i_2} \rangle$, let $\mathbf{x}_i$ be the $(d + 1)$-dimensional vector including constant 1 and d topological features between them, and y_i be the label of whether they will be co-authors in the future ($y_i = 1$ if they will be co-authors, and 0 otherwise), which follows Bernoulli distribution with probability p_i ($P(y_i = 1) = p_i$). The probability p_i is modeled as follows:

$$p_i = \frac{e^{\mathbf{x}_i \beta}}{e^{\mathbf{x}_i \beta} + 1},$$

where β is the $d + 1$ coefficient weights associated with the constant and each topological feature. We then use standard MLE (Maximum Likelihood Estimation) to derive $\hat{\beta}$, that maximizes the likelihood of all the training pairs:

$$L = \prod_i p_i^{y_i} (1 - p_i)^{(1 - y_i)}.$$

In the test stage, for each candidate author pair, we can predict whether they will collaborate according to $P(y_{test} = 1) = \dfrac{e^{x_{test}\hat{\beta}}}{e^{x_{test}\hat{\beta}}+1}$, where $\mathbf{x}_{test}$ is the $(d + 1)$-dimensional vector including constant 1 and d topological features between the candidate pair.

5.3.2 EXPERIMENTAL RESULTS

It turns out that the proposed meta-path-based topological features can improve the co-authorship prediction accuracy compared with the baselines that only use homogeneous object and link information.

We consider three time intervals for the DBLP network, according to the publication year associated with each paper: T_0 = [1989, 1995], T_1 = [1996, 2002], and T_2 = [2003, 2009]. For the training stage, we use T_0 as the past time interval, and T_1 as the future time interval, which is denoted as $T_0 - T_1$ time framework. For the test stage, we consider the same time framework $T_0 - T_1$ for most of the studies, and consider $T_1 - T_2$ time framework for the query-based case study.

Let an author pair be $\langle a_i, a_j \rangle$, we call a_i the source author, and a_j the target author. Two sets of source authors are considered. The first set is comprised of highly productive authors, who has published no less than 16 papers in the past time interval; and the second set is comprised of less productive authors, with between 5 and 15 publications. The target authors are selected if they are 2-hop co-authors or 3-hop co-authors of a source author. In all, we have four labeled datasets: (1) the highly productive source authors with 2-hop target authors (denoted as *H P2hop*); (2) the highly productive source authors with 3-hop target authors (denoted as *H P3hop*); (3) the less productive source authors with 2-hop target authors (denoted as *L P2hop*); and (4) the less productive source authors with 3-hop target authors (denoted as *L P3hop*).

To evaluate the prediction accuracy, two measures are used. The first measure is the classification accuracy rate (accuracy) for binary prediction under the cut-off score as 0.5, and the second one is the area under ROC (receiver operating characteristic) curve [9], which is denoted as AUC.

Overall Accuracy

We first compare the heterogeneous topological features with the homogeneous ones. For the heterogeneous topological features, we use path count measure for 9 meta-paths (denoted as heterogeneous PC) listed in Table 5.1 (not including the target relation itself); for homogeneous topological features, we use: (1) the number of common co-authors; (2) the rooted PageRank [39] with restart probability $\alpha = 0.2$ for the induced co-author network; and (3) the number of paths between two authors of length no longer than 4, disregarding their different meta-paths (denoted as homogeneous PC). The rooted PageRank measure is only calculated for the $HP3hop$ dataset, due to its inefficiency in calculation for large number of authors. The comparison results are summarized in Figure 5.3 and Table 5.2. We can see that the heterogeneous topological feature beats the homogeneous ones in all the four datasets, which validates the necessity to consider the different meta-paths separately in heterogeneous networks. We also notice that, in general, the co-authorship for highly productive authors is easier to predict than less productive authors, by looking at the overall prediction accuracy on the two groups of source authors. Finally, we can see that the prediction accuracy is higher when the target authors are 3-hop co-authors, which means the collaboration between closer authors in the network is more affected by information that is not available from network topology.

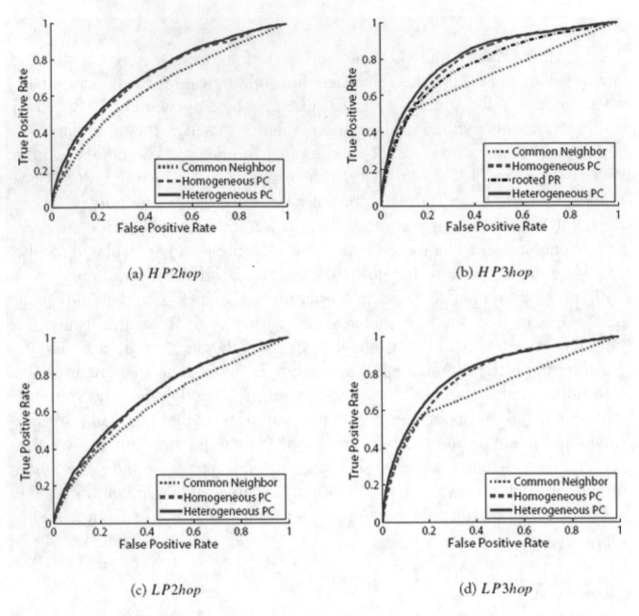

(a) *H P2hop*

(b) *H P3hop*

(c) *LP2hop*

(d) *LP3hop*

Figure 5.3: Homogeneous features vs. heterogeneous *Path Count* feature. Heterogeneous feature beats homogeneous features in all of the datasets, which is more significant on 3-hop datasets, where topological features play a more important role for co-authorship prediction.

Table 5.2: Homogeneous vs. heterogeneous topological features

Dataset	Topological features	Accuracy	AUC
H P2hop	common neighbor	0.6053	0.6537
	homogeneous PC	0.6433	0.7098
	heterogeneous PC	**0.6545**	**0.7230**
H P3hop	common neighbor	0.6589	0.7078
	homogeneous PC	0.6990	0.7998
	rooted PageRank	0.6433	0.7098
	heterogeneous PC	**0.7173**	**0.8158**
L P2hop	common neighbor	0.5995	0.6415
	homogeneous PC	0.6154	0.6868
	heterogeneous PC	**0.6300**	**0.6935**
L P3hop	common neighbor	0.6804	0.7195
	homogeneous PC	0.6901	0.7883
	heterogeneous PC	**0.7147**	**0.8046**

Second, we compare different measures proposed for heterogeneous topological features: (1) the path count (*PC*); (2) the normalized path count (*N P C*, i.e., PathSim in our case); (3) the random walk (*RW*); (4) the symmetric random walk (*S R W*); and (5) the hybrid features of (1)–(4) (*hybrid*). It turns out that in general we have (see Figure 5.4): (1) all the heterogeneous features beat the homogeneous features (common neighbor is denoted as *PC*1, and homogeneous PC is denoted as *P C Sum*); (2) the normalized path count beats all the other three individual measures; and (3) the hybrid feature produces the best prediction accuracy.

Case Study

For the case study, we first show the learned importance for each topological feature in deciding the relationship building in DBLP, and then show the predicted co-author relationships for some source author in a query mode.

First, we show the learned importance for all the 9 meta-paths with *N P C* measure, as *N P C* is the best measure for co-author relationship prediction overall. We show the *p*-value for the feature associated with each meta-path under Wald test and their significance level in Table 5.3. From the results, we can see that for the *H P3hop* dataset, the shared co-authors, shared venues, shared topics, and co-cited papers for two authors all play very significant roles in determining their future collaboration(s). For the asymmetric meta-paths that represent the asymmetric relations, such as citing and cited relations between authors, they have different impacts in determining the relationship building. For example, for a highly productive source author, the target authors citing her frequently are more likely to be her future co-authors than the target authors being cited by her frequently.

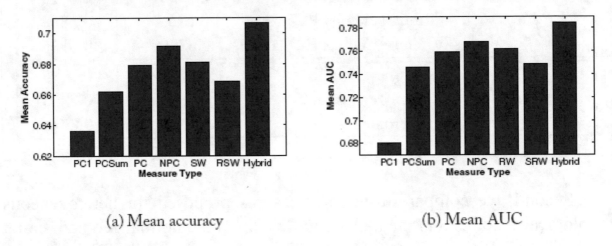

(a) Mean accuracy   (b) Mean AUC

Figure 5.4: Average accuracy over four datasets for different features.

Table 5.3: Significance of meta-paths with *Normalized Path Count measure* for *HP 3hop* dataset

Meta-path	p-value	Significance level[1]
$A - P \to P - A$	0.0378	**
$A - P \leftarrow P - A$	0.0077	***
$A - P - V - P - A$	1.2974e-174	****
$A - P - A - P - A$	1.1484e-126	****
$A - P - T - P - A$	3.4867e-51	****
$A - P \to P \to P - A$	0.7459	
$A - P \leftarrow P \leftarrow P - A$	0.0647	*
$A - P \to P \leftarrow P - A$	9.7641e-11	****
$A - P \leftarrow P \to P - A$	0.0966	*

[1] *: $p < 0.1$; **: $p < 0.05$; ***: $p < 0.01$, ****: $p < 0.001$

Second, we study the predicted co-authors for some source author as queries. Note that, predicting co-authors for a given author is an extremely difficult task, as we have too many candidate target authors (3-hop candidates are used), while the number of real new relationships are usually quite small. Table 5.4 shows the top-5 predicted co-authors in time interval T_2 (2003–2009) using the $T_0 - T_1$ training framework, for both the proposed hybrid topological features and the shared co-author feature. We can see that the results generated by heterogeneous features has a higher accuracy compared with the homogeneous one.

Table 5.4: Top-5 predicted co-authors for Jian Pei in 2003–2009

Rank	Hybrid heterogeneous features	# of shared authors as features
1	**Philip S. Yu**	Philip S. Yu
2	**Raymond T. Ng**	Ming-Syan Chen
3	Osmar R. Zaïane	Divesh Srivastava
4	**Ling Feng**	Kotagiri Ramamohanarao
5	**David Wai-Lok Cheung**	Jeffrey Xu Yu

* Bold font indicates true new co-authors of Jian Pei in the period of 2003-2009.

5.4 RELATIONSHIP PREDICTION WITH TIME

Traditional link prediction studies have been focused on asking **whether** a link will be built in the future, such as "whether two people will become friends?" However, in many applications, it may be more interesting to predict **when** the link will be built, such as "what is the probability that two authors will co-write a paper within 5 years?" and "by when will a user in Netflix rent the movie *Avatar* with 80% probability?"

In this section, we study the problem of predicting the *relationship building time* between two objects, such as when two authors will collaborate for the first in the future, based on the topological structure in a heterogeneous network, by investigating the citation relationship between authors in the DBLP network. First, we introduce the concepts of target relation and topological features for the problem encoded in meta-paths [65]. Then, a generalized linear model (GLM) [19] based supervised framework is proposed to model the relationship building time. In this framework, the building time for relationships are treated as independent random variables conditional on their topological features, and their expectation is modeled as a function of a linear predictor of the extracted topological features. We propose and compare models with different distribution assumptions for relationship building time, where the parameters for each model are learned separately.

5.4.1 META-PATH-BASED TOPOLOGICAL FEATURES FOR AUTHOR CITATION RELATIONSHIP PREDICTION

In the author citation relationship prediction problem, the target relation is $A - P \rightarrow P - A$, which is short for $A \xrightarrow{write} P \xrightarrow{cite} P \xrightarrow{write^{-1}} A$, and describes the citation relation between authors. In general, for a target relation $R_T = \langle A, B \rangle$, any meta-paths starting with type A and ending with type B other than the target relation itself can be used as the topological features for predicting new relationships. These meta-paths can be obtained by traversing on the network schema, for example, using BFS (breadth-first search). By reasoning the dynamics of a relationship building, we are in particular considering three forms of relations as topological features.

Table 5.5: Meta-paths denoting similarity relations between authors

Meta-path	Semantic Meaning of the Relation
$A - P - A$	a_i and a_j are co-authors
$A - P - V - P - A$	a_i and a_j publish in the same venues
$A - P - A - P - A$	a_i and a_j are co-authors of the same authors
$A - P - T - P - A$	a_i and a_j write the same topics
$A - P \rightarrow P \leftarrow P - A$	a_i and a_j cite the same papers
$A - P \leftarrow P \rightarrow P - A$	a_i and a_j are cited by the same papers

1. $A\ R_{sim}\ A\ R_T B$, where R_{sim} is a similarity relation defined between type A and R_T is the target relation. The intuition is that if a_i in type A is similar to many a_k's in type A that have relationships with b_j in type B, then a_i is likely to build a relationship with b_j in the future.

2. $A\ R_T B\ R_{sim}\ B$, where R_T is the target relation, and R_{sim} is a similarity relation between type B. The intuition is that if a_i in type A has relationships with many b_k's in type B that are similar to b_j in type B, then a_i is likely to build a relationship with b_j in the future.

3. $A\ R_1 C\ R_2 B$, where R_1 is some relation between A and C and R_2 is some relation between C and B. The intuition is that if a_i in type A has relationships with many c_k's in type C that have relationships with b_j in type B, then a_i is likely to build a relationship with b_j in the future. Note that the previous two forms are special cases of this one, which can be viewed as triangle connectivity property.

For topological features, we confine similarity relations R_{sim} and other partial relations R_1 and R_2 to those that can be derived from the network using meta-paths. Moreover, we only consider similarity relations that are symmetric.

Taking the author citation relation, which is defined as $A - P \rightarrow P - A$, as the target relation, we consider 6 author-author similarity relations defined in Table 5.5. For each similarity relation, we can concatenate the target relation in its left side or in its right side. We then have 12 topology features with the form $A\,R_{sim}\,A\,R_T\,B$ and $A\,R_T\,B\,R_{sim}\,B$ in total. Besides, we consider the concatenation of "author-cites-paper" relation ($A - P \rightarrow P$) and "paper-cites-author" relation ($P \rightarrow P - A$) into ($A - P \rightarrow P \rightarrow P - A$), as well as all the 6 similarity relations listed in Table 5.5, which can be viewed as the form of $A\,R_1\,C\,R_2\,B$ themselves. Now we have 19 topological features in total.

For each type of the meta-paths, we illustrate a concrete example to show the possible relationship building in Figure 5.5. In Figure 5.5(a), authors a_1 and a_2 are similar, as they publish papers containing similar terms, and a_2 cites papers published by a_3. In the future, a_1 is likely to cite papers published by a_3 as well, since she may follow the behavior of her fellows. In Figure 5.5(b), author a_1 cites a_2, and a_2 and a_3 are cited by common papers together (p_5, p_6, p_7). Then a_1 is likely to cite a_3 in the future, as she may cite authors similar to a_2. In Figure 5.5(c), a_1 and a_2 publish in the same venue, then a_1 is likely to cite a_2 in the future as they may share similar interests if publishing in the same conference.

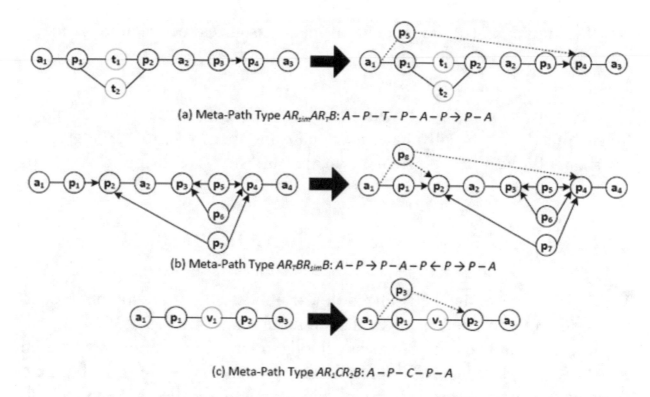

(a) Meta-Path Type $AR_{sim}AR_7B$: $A - P - T - P - A - P \rightarrow P - A$

(b) Meta-Path Type $AR_7BR_{sim}B$: $A - P \rightarrow P - A - P \leftarrow P \rightarrow P - A$

(c) Meta-Path Type AR_1CR_2B: $A - P - C - P - A$

Figure 5.5: Feature meta-path illustration for author citation relationship prediction.

By varying the similarity relations and partial relations, we are able to generate other topological features in arbitrary heterogeneous networks.

Without loss of generality, we use the count of path instances as the default measure. Thus, each meta-path corresponds to a measure matrix. For a single relation $R \in \mathcal{R}$, the measure matrix is just the adjacency matrix of the subnetwork extracted by R. Given a composite relation, the measure matrix can be calculated by the matrix multiplication of the partial relations.

In Figure 5.5(a), the count of path instances between a_1 and a_3 following the given meta-path is 2, which are:

(1) $a_1 - p_1 - t_1 - p_2 - a_2 - p_3 \rightarrow p_4 - a_3$, and

(2) $a_1 - p_1 - t_2 - p_2 - a_2 - p_3 \rightarrow p_4 - a_3$.

In Figure 5.5(b), the count of path instances between a_1 and a_4 following the given meta-path is 3, which are:

(1) $a_1 - p_1 \rightarrow p_2 - a_2 - p_3 \leftarrow p_5 \rightarrow p_4 - a_4$,

(2) $a_1 - p_1 \rightarrow p_2 - a_2 - p_3 \leftarrow p_6 \rightarrow p_4 - a_4$, and

(3) $a_1 - p_1 \rightarrow p_2 - a_2 - p_2 \leftarrow p_7 \rightarrow p_4 - a_4$.

In Figure 5.5(c), the count of path instances between a_1 and a_3 following the given meta-path is 1, which is:

(1) $a_1 - p_1 - v_1 - p_2 - a_3$.

Measures for different meta-paths have different scales. For example, longer meta-paths usually have more path instances due to the adjacency matrix multiplication. We will normalize the measure using Z-score for each meta-path.

5.4.2 THE RELATIONSHIP BUILDING TIME PREDICTION MODEL

We now propose the generalized linear model-based prediction model, which directly models the relationship building time as a function of topological features, and provides methods to learn the coefficients of each topological feature, under different assumptions for relationship building time distributions. After that, we introduce how to use the learned model to make inferences.

We model the relationship building time prediction problem in a supervised learning framework. In the **training stage**, we first collect the topological features $\mathbf{x}_i$ in the history interval $T_0 = [t_0, t_1)$ for each sampled object pair $\langle a_i, b_i \rangle$, where types of a_i and b_i are $\tau(a_i) = A$ and $\tau(b_i) = B$. Then, we record their relative first relationship building time $y_i = t_i - t_1$, if t_i is in the future training interval $T_1 = [t_1, t_2)$; record the building time $y_i \geq t_2 - t_1$, if no new relationship has been observed in T_1. Note that in the training stage, we are only given limited time to observe whether and when two objects will build their relationship, it is very possible that two objects build their relationship after t_2, which needs careful handling in the training model. A generalized linear model (GLM) based relationship building time model is introduced, and the goal is to learn the best coefficients associated with each topological feature that maximize the current observations of the relationship building time. In the **test stage**, we apply the learned coefficients of the topological features to the test pairs, and compare the predicted relationship building time with the ground truth.

Different from the existing link prediction task, in the training stage, we are collecting relationship building time y_i for each training pair, which is a variable ranging from 0 to ∞, rather than a binary value denoting whether there exists a link in the future interval. Similarly, in the test stage, we are predicting the relationship building time y_i for test pairs that range from 0 to ∞, rather than predicting whether the link exists or not in the given future interval.

The Generalized Linear Model Framework

The main idea of generalized linear model (GLM) [19] is to model the expectation of random variable Y, $E(Y)$, as some function ("link function") of the linear combination of features, that is, $\mathbf{X}\boldsymbol{\beta}$, where $\mathbf{X}$ is the observed feature vector, and $\boldsymbol{\beta}$ is the coefficient vector. Then the goal is to learn $\boldsymbol{\beta}$ according to the training data set using maximum likelihood estimation. Under different distribution assumptions for Y, usually from the exponential family, $E(Y)$ has different forms of parameter set, and the link functions are with different forms too. Note that the most frequently used Least-Square regression and logistic regression are special cases of GLM, where Y follows Gaussian distribution and Bernoulli distribution, respectively.

Suppose we have n training pairs for the target relation $\langle A, B \rangle$. We denote each labeled pair as $r_i = \langle a_i, b_i \rangle$, and y_i as the observed relative relationship building time in the future interval. We denote $\mathbf{X}_i$ as the d dimensional topological feature vector extracted for a_i and b_i in the historical interval plus a constant dimension.

Distributions for Relationship Building Time

The first issue of the prediction model is to select a suitable distribution for the relationship building time. Intuitively, a relationship building between two objects can be treated as an event, and we are interested in when this event will happen.

Let Y be the relationship building time relative to the beginning of the future interval ($y_i = t_i - t_1$), and let T be the length of future training interval. For training pairs, Y has the observations in $[0,T) \cup \{T^+\}$ in a continuous case, and $\{0, 1, 2, \ldots, T-1, T^+\}$ in a discrete case, where $y = T^+$ means no event happens within the future training interval. For testing pairs, Y has the observations in $[0, \infty)$ in a continuous case, and nonnegative integers in a discrete case.

We consider three types of distributions for relationship building time, namely exponential, Weibull and geometric distribution. For each of the distribution assumptions over y_i, we set up the models separately.

The first distribution is **exponential distribution**, which is the most frequently used distribution in modeling waiting time for an event. The probability density function of an exponential distribution is:

$$f_Y(y) = \frac{1}{\theta} \exp\{-\frac{y}{\theta}\} , \qquad (5.1)$$

where $y \geq 0$, and $\theta > 0$ is the parameter denoting the *mean waiting time* for the event. The cumulative distribution function is:

$$F_Y(y) = Pr(Y \leq y) = 1 - \exp\{-\frac{y}{\theta}\} . \qquad (5.2)$$

The second distribution is **Weibull distribution**, which is a generalized version of exponential distribution and is another standard way to model the waiting time of an event. The probability density function of a Weibull distribution is:

$$f_Y(y) = \frac{\lambda y^{\lambda-1}}{\theta^{\lambda}} \exp\{-(\frac{y}{\theta})^{\lambda}\} , \qquad (5.3)$$

where $y \geq 0$, and $\theta > 0$ and $\lambda > 0$ are two parameters related to *mean waiting time* for the event and *hazard of happening* of the event along with the time. λ is also called the shape parameter, as it affects the shape of probability function. When $\lambda > 1$, it indicates an increasing happening rate along the time (if an event does not happen at an early time, it is getting higher probability to happen at later time); and when $\lambda < 1$, it indicates a decreasing happening rate along the time (if an event does not happen at an early time, it is getting less possible in happening in later time). Note that when $\lambda = 1$, Weibull distribution becomes exponential distribution with mean waiting time as θ, and the happening rate does not change along the time. The cumulative distribution function is:

$$F_Y(y) = Pr(Y \le y) = 1 - \exp\{-(\frac{y}{\theta})^\lambda\}. \tag{5.4}$$

The third distribution is the **geometric distribution**, which is a distribution that models how many times of failures it needs to take before the first-time success. As in our case, the time of failure is the discrete time that we need to wait before a relationship is built. The probability mass function of a geometric distribution is:

$$Pr(Y = k) = (1 - p)^k p, \tag{5.5}$$

where $k = 0, 1, 2, \ldots$, and p is the probability of the occurrence of the event at each discrete time. The cumulative distribution function is:

$$Pr(Y \le k) = 1 - (1 - p)^{k+1}. \tag{5.6}$$

In our case, each relationship building is an independent event, and each relationship building time Y_i is an independent random variable, following the same distribution family, but with different parameters. With the distribution assumptions, we build relationship building time prediction models in the following.

Model under Exponential and Weibull Distribution Note that, as exponential distribution is a special case of Weibull distribution (with $\lambda = 1$), we only discuss prediction model with Weibull distribution.

In this case, we assume relationship building time Y_i for each training pair is independent of each other, following the same Weibull distribution family with the same shape parameter λ, but with different mean waiting time parameters θ_i. Namely, we assume that different relationships for the target relation share the same trend of hazard happening along with the time, but with different expectation in building time. Under this assumption, we can evaluate the expectation for each random variable Y_i as $E(Y_i) = \theta_i \Gamma(1 + \frac{1}{\lambda})$. We then use the link function $E(Y_i) = \exp\{-\mathbf{X}_i\boldsymbol{\beta}\}\Gamma(1 + \frac{1}{\lambda})$, that is

$\log \theta_i = -\beta_0 - \sum_{j=1}^{d} X_{i,j}\beta_j = -X_i\beta$, where β_0 is the constant term. Then we can write the log-likelihood function:

$$\log L = \sum_{i=1}^{n} (f_Y(y_i|\theta_i, \lambda)I_{\{y_i < T\}} + P(y_i \geq T|\theta_i, \lambda)I_{\{y_i \geq T\}}) ,$$

where $I_{\{y_i < T\}}$ and $I_{\{y_i \geq T\}}$ are indicator functions, which equals to 1 if the predicate holds, or 0 otherwise. It is easy to see that the log-likelihood function includes two parts: if y_i is observed in the future interval, we use its real density in the function; otherwise, we are only able to use the probability of $y_i \geq T$ in the function.

By plugging in $\log \theta_i = -X_i\beta$, we can get the log-likelihood with parameters β and λ:

$$LL_W(\beta, \lambda) = \sum_{i=1}^{n} I_{\{y_i < T\}} \log \frac{\lambda y_i^{\lambda-1}}{e^{-\lambda X_i\beta}} - \sum_{i=1}^{n} \left(\frac{y_i}{e^{-X_i\beta}}\right)^\lambda , \tag{5.7}$$

where LL_W denotes the **log-likelihood** function under **Weibull** distribution. We refer this model as Weibull model.

Model under Geometric Distribution In this case, we assume relationship building time Y_i for each training pair is independent of each other, following the same geometric distribution family, but with different success probability p_i. Under this assumption, we can evaluate the expectation for each random variable Y_i as $E(Y_i) = \frac{1-p_i}{p_i}$. We then let $E(Y_i) = \exp\{-X_i\beta\}$, i.e., $\log\frac{1-p_i}{p_i} = -X_i\beta$. The log-likelihood function is then:

$$LL_G(\beta) = \sum_{i=1}^{n} (Pr(Y_i = y_i)I_{\{y_i < T\}} + P(y_i \geq T)I_{\{y_i \geq T\}})$$
$$= \sum_{i=1}^{n} \left(-I_{\{y_i < T\}}(-X_i\beta) + (y_i + 1)(-X_i\beta - \log(e^{-X_i\beta} + 1))\right) . \tag{5.8}$$

We refer this model as geometric model.

The learning of the models is becoming an optimization problem, which aims at finding $\hat{\beta}$ and other parameters (e.g., $\hat{\lambda}$ in the Weibull model) that maximize the log-likelihood. As there are no closed form solutions for Equations (5.7) and (5.8), we use standard Newton-Raphson method to derive the update formulas, which are based on the first derivative and second derivative (Hessian matrix) of the log-likelihood function.

Model Inference

Once the parameters such as β and λ are learned from the training data set through MLE, we can apply the model to the test pairs of objects, as long as their topological features in the historical network are given. Let the learned parameter values be $\hat{\beta}$ and $\hat{\lambda}$ for β and λ, and let the topological feature vector for the test pairs be $\mathbf{X}_{test}$ (with constant 1 as the first dimension), we now consider three types of questions people may be interested in for the new relationship building time, and provide the solutions in the following.

1. Will a new relationship between two test objects be built within t years?

   This question is equal to the query for the probability $Pr(y_{test} \leq t)$, which can be evaluated by plugging in the MLE estimators to derive the distribution parameters. Note that for traditional link prediction tasks, t should be the same as the length of training interval. For our task, t can be any nonnegative values. For Weibull model, we have:

$$\hat{\theta}_{test} = \exp\{-\mathbf{X}_{test}\hat{\beta}\}$$
$$Pr(y_{test} \leq t) = 1 - \exp\{-(\frac{t}{\hat{\theta}_{test}})^{\hat{\lambda}}\}. \tag{5.9}$$

   For geometric model, we have:

$$\hat{p}_{test} = \frac{1}{\exp\{-\mathbf{X}_{test}\hat{\beta}\} + 1}$$
$$Pr(y_{test} \leq t) = 1 - (1 - \hat{p}_{test})^{t+1}. \tag{5.10}$$

2. What is the average relationship building time for two test objects?

This is simply the query for $E(Y_{test})$. Using the same estimators for $\hat{\theta}_{test}$ and $\hat{P}_{test}$ as above, we can have the estimator for $E(Y_{test})$ as $E(Y_{test}) = \hat{\theta}_{test}$ $\Gamma(1 + \frac{1}{\hat{\lambda}})$ for Weibull model, where $\Gamma(\cdot)$ is the Gamma function, and $E(Y_{test}) = \frac{1 - \hat{p}_{test}}{\hat{p}_{test}}$ for geometric model.

3. The quantile: by when a relationship will be built with a probability α?

This is equal to query for the solution of $F_Y(y_{test}) = \alpha$, and we can get answers as $y_{test} = \hat{\theta}_{test}(-\log(1-\alpha))^{\frac{1}{\hat{\lambda}}}$ for Weibull model, and $y_{test} = \max\max\{\frac{\log(1-\alpha)}{\log(1-\hat{p}_{test})} - 1, 0\}$ for geometric model. When $\alpha = 0.5$, the quantile is just the median.

5.4.3 EXPERIMENTAL RESULTS

We select a subset of authors in the DBLP bibliographic network, who published more than 5 papers in top conferences in the four areas[1] that are related to data mining between years 1996 and 2000 ($T_0 = [1996, 2000]$). The total number of the author set is 2721. Then we sampled 7000 pairs of authors in the form of $\langle a_i, a_j \rangle$ that a_i did not cite a_j in T_0, but have citation relationship between year 2001 and 2009 ($T_1 = [2001, 2009]$ and $T = 9$) as positive samples; and we sampled another 7000 pairs of authors that have no citation relationship during either T_0 or T_1. The citation relationship is defined if a_i cites papers written by a_j published before year 2000. Note that we have this time constraint for papers as we want to infer citation relationship via the historical network. Nineteen topological features introduced in Section 5.4.1 are calculated for each training pair. The first (relative) time of the citation relationship is recorded for each pair of authors; and if there is no citation relationship between them in T_1, the time is recorded as a value bigger than 9.

Experimental Setting

In order to show the power of using time-involved model in relationship prediction, we use logistic regression [52] (denoted as *logistic*) that is frequently used in binary link prediction tasks as the baseline. Note that the output of the logistic regression is a probability denoting whether a relationship will be built in T_1 for each test pair. In our models, the output is the parameter set for the distribution of the relationship building time, from which we can infer much more information rather than a simple probability. We denote our models with different distribution assumptions as *GLM_geo*, *GLM_exp*, and *GLM_weib*, respectively.

To compare the four models, we use two sets of measures to evaluate the effectiveness of each model. First, we measure the effectiveness according to the predicted probability for each relationship. We define the *accuracy* of the relationship prediction as the ratio between the number of correctly predicted relationship (under the *cut-off* 0.5) and the total number of the test pairs. Also, another frequently used measure AUC (the area under ROC curve) is used to compare the accuracy.

Second, we directly compare the predicted time with the ground truth, among our proposed models. Mean absolute error (*MAE*) that is the mean of the absolute error between predicted relationship building time and the ground truth is used. Also, we use the ratio of the relationships that occur in some confidence interval derived from the models as another measure to test the accuracy of the predicted time. Note that relationships yet to happen are not considered in these two measures.

Prediction Power Study

We now compare our time-involved models with the baseline logistic regression, using the first set of measures.

We test the generality power for different models, namely, when the training future interval is not equal to the test future interval ($T^{train} \neq T^{test}$). On one hand, we may want to know the probability of relationship building within each year in the training interval ($T^{test} < T^{train}$); on the other hand, we may want to infer longer term probability given a short term training interval ($T^{test} > T^{train}$). We show the two cases in Tables 5.6 and 5.7. Note that since logistic regression can only output the probability when $T^{test} = T^{train}$, we use the same predicted probability for different test intervals. In Table 5.6, we fix the training interval with length $T^{train} = 9$, namely, T_1^{train}

= [2001, 2009], and vary the test intervals with length from 1–4. We can see that when T^{test} is small, time-involved models can give much better prediction accuracy, especially in terms of the measure *accuracy*. In other words, time-involved models carry more information in telling the probability of relationship building in finer time periods. In Table 5.7, we fix the test interval with length $T^{test} = 9$ and vary the training intervals with length from 2–5. We can see that time-involved models can better utilize the short-term training than logistic regression, and output better prediction results for longer term relationship building behavior. It is interesting to note that by using the measure *AUC*, which does not require users to specify a *cut-off* value in the predicted probabilities, the performance of logistic regression is still comparable with other models. This is due to *AUC* only uses the ranking order of the predicted values, while *accuracy* requires that the *absolute values* of the predicted probabilities are also correct.

Table 5.6: Prediction generalization power comparison: $T^{test} < T^{train}$ and $T^{train} = 9$

	$T^{test} = 1$		$T^{test} = 2$		$T^{test} = 3$		$T^{test} = 4$	
	Accuracy	AUC	Accuracy	AUC	Accuracy	AUC	Accuracy	AUC
logistic	0.7106	0.7619	0.7246	**0.7535**	0.7669	0.7347	**0.7349**	**0.7731**
GLM-geo	0.9284	**0.7626**	0.8436	0.7532	**0.7829**	**0.7657**	0.7347	0.7696
GLM-exp	**0.9290**	0.7553	**0.8442**	0.7464	0.7821	0.7569	0.7328	0.7603
GLM-weib	0.9287	0.7273	0.8441	0.7452	0.7826	0.7559	0.7334	0.7597

In all, for time-involved model, it contains more information and can answer different questions and with strong generalization power. Logistic regression can only answer the question of whether a relationship will happen or not, given a fixed time interval. However, if we are asking more, it fails in most of the scenarios.

Table 5.7: Prediction generalization power comparison: $T^{test} > T^{train}$ and $T^{test} = 9$

	$T^{train}=2$		$T^{train}=3$		$T^{train}=4$		$T^{train}=5$	
	Accuracy	AUC	Accuracy	AUC	Accuracy	AUC	Accuracy	AUC
logistic	0.5157	0.7810	0.5379	0.7805	0.5599	0.7841	0.5952	0.7896
GLM-geo	0.5942	**0.7910**	0.6209	**0.7926**	0.6366	**0.7902**	0.6522	**0.7982**
GLM-exp	0.5015	0.7802	0.5214	0.7833	0.6709	0.7841	**0.7143**	0.7870
GLM-weib	**0.7081**	0.7816	**0.7021**	0.7832	**0.7002**	0.7833	0.7103	0.7862

Time Prediction Accuracy Study

We now evaluate the predicted time using different time-involved models. Here, we use the predicted median time as the predicted time. Table 5.8 shows the MAE (mean average error) between the predicted median time and the ground truth under different training and test intervals. It turns out that *GLM-exp* has the lowest error. Also, both *GLM-exp* and *GLM-weib* perform even better using shorter interval as training, whereas *GLM-geo* has the opposite behavior, that is, longer term of training leads to better performance. Note that we only calculate the error for the relationships indeed happen in the test interval.

In Table 5.9, we infer different confidence intervals from the predicted relationship building time distribution, and test the ratio of the true relationship in different confidence intervals. A confidence interval (range) rather than a simple value, say the median time, can give users a better view of the relationship building time. It is shown that *GLM-exp* and *GLM-weib* has a higher ratio of giving correct confidence intervals for the true relationship building time, especially when using a small confidence interval. This is very useful in practice as they can give tight bound estimations.

Table 5.8: *MAE* of predicted time and the ground truth

	$T^{train}=5, T^{test}=9$	$T^{train}=9, T^{test}=9$
GLM-geo	4.9883	4.7219
GLM-exp	2.7774	3.0685
GLM-weib	3.1025	3.1692

Case Studies

To better understand the output of our model, we now show a case study of predicting when the citation relationship will be build for "Philip S. Yu" with other candidates. The model is trained by *GLM-weib* using a training interval of 9 years (T_1^{train} = [2001, 2009]), with the learned parameter λ = 0.9331, slightly less than 1, which means the citation relationship has a higher hazard happening at an earlier time. The ground truth of the citation building time, and the predicted median, mean, 25% quantile and 75% quantile for several test pairs are shown in Table 5.10. It can be seen that the predicted median and confidence interval are very suggestive for predicting the true citation relationship building time. For those authors whose predicted being cited time is significantly different from the ground truth, in-depth studies may be needed. For example, David Maier is a prolific researcher in database area, and by intuition as well as suggested by the model, Philip should cite him. However, the ground truth says otherwise. Furthermore, this function can be used to recommend authors to any author in DBLP for citation purpose.

Table 5.9: Ratio of the true relationship occurring in different confidence intervals: T^{test} = 9

	25%-75%		10%-90%		0%-80%	
	$T^{train} = 9$	$T^{train} = 5$	$T^{train} = 9$	$T^{train} = 5$	$T^{train} = 9$	$T^{train} = 5$
GLM-geo	0.5489	0.5336	**0.8936**	**0.8947**	0.9650	0.9743
GLM-exp	0.7167	0.7246	0.8619	0.8634	0.9880	0.9889
GLM-weib	**0.7278**	**0.7314**	0.8680	0.8686	**0.9884**	**0.9896**

Table 5.10: Case studies of relationship building time prediction

a_i	a_j	Ground Truth	Median	Mean	25% quant.	75% quant.
Philip S. Yu	Ling Liu	1	2.2386	3.4511	0.8549	4.7370
Philip S. Yu	Christian Jensen	3	2.7840	4.2919	1.0757	5.8911
Philip S. Yu	C. Lee Giles	0	8.3985	12.9474	3.2450	17.7717
Philip S. Yu	Stefano Ceri	0	0.5729	0.8833	0.2214	1.2124
Philip S. Yu	David Maier	9+	2.5675	3.9581	0.9920	5.4329
Philip S. Yu	Tong Zhang	9+	9.5371	14.7028	3.6849	20.1811
Philip S. Yu	Rudi Studer	9+	9.7752	15.0698	3.7769	20.6849

For the above model, the learned *top-4* most important topological features with the highest coefficients are:

1. $A - P - T - P - A$, that is, if two authors are very similar in terms of writing similar topics, they tend to cite each other;

2. $A - P \leftarrow P \rightarrow P - A$, that is, if two authors are very similar in terms of being frequently co-cited by the common papers, they tend to cite each other;

3. $A - P - A - P \rightarrow P - A$, that is, an author tends to cite the authors that are frequently cited by her co-authors; and

4. $A - P - T - P - A - P \rightarrow P - A$, that is, if two authors are similar in terms writing similar topics, they tend to cite the same authors.

These topological features provide insightful knowledge for people in understanding the citation relationship building between authors.

---

[1]Data Mining: KDD, PKDD, ICDM, SDM, PAKDD; Database: SIGMOD Conference, VLDB, ICDE, PODS, EDBT; Information Retrieval: SIGIR, ECIR, ACL, WWW, CIKM; and Artificial Intelligence: NIPS, ICML, ECML, AAAI, IJCAI.

PART III
Relation Strength-Aware Mining

Relation Strength-Aware Clustering with Incomplete Attributes

A heterogeneous information network contains multiple types of objects as well as multiple types of links, indicating different sorts of interactions among these objects. The heterogeneity of network model brings rich semantic information for mining. It also raises the issue of selecting the right type of information for different mining purposes. For mining different kinds of knowledge, it is desirable to automatically learn the right information encoded in the network, with limited guidance from users. In this chapter, we study a special case of such problems: cluster objects in a network, with user-provided attribute set and relations from the original network schema.

6.1 OVERVIEW

The rapid emergence of online social media, e-commerce, and cyber-physical systems brings the necessity to study them with the model of heterogeneous networks in which objects (i.e., nodes) are of different types, and links among objects correspond to different relations, denoting different interaction semantics. In addition, an object is usually associated with some attributes. For example, in a *YouTube* social media network, the object types may include videos, users, and comments; links between objects correspond to different relations, such as publish and like relations between users and videos, post relation between users and comments, and friendship and subscribe relations between users; and attributes may include user's location, the length of video's clips, the number of views, and comments.

Such kinds of heterogeneous information networks are ubiquitous and determining their underlying clustering structures has many interesting applications. For example, clustering objects (e.g., customers, products, and comments) in an online shopping network such as *eBay* is helpful for customer segmentation in product marketing; and clustering objects (e.g., people, groups, books, and posts) in an online social network such as *Facebook* is helpful for voter segmentation in political campaigns.

The clustering task brings two new challenges in such scenarios. First, an object may contain only partial or even no observations for a given attribute set that is critical to determine their cluster labels. That is, a pure attribute-based clustering algorithm cannot correctly detect these clusters. Second, although links have been frequently used in networks to detect clusters [2; 14; 44; 69] in recent research, we consider a more challenging scenario in which the links are of different types and interpretations, each of which may have its own level of semantic importance in the clustering process. *That is, a pure link-based clustering without any guidance from attribute specification could fail to meet user demands.*

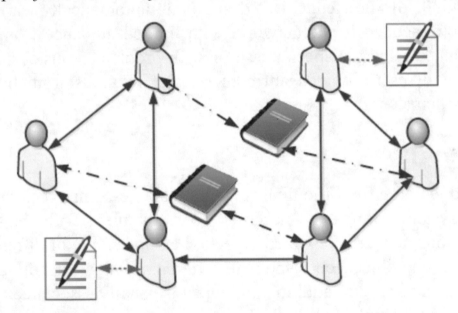

Figure 6.1: A motivating example on clustering political interests in social information networks.

Figure 6.1 shows a toy social information network extracted from a political forum containing users, blogs written by users, books liked by users, and friendship between users. Now suppose we want to cluster users in the network according to their political interests, using the text attributes in user profiles, blogs and books, as well as the link information between objects. On one hand, since not all the users listed their political interests in their profiles, we cannot judge their political interests simply according to the text information contained in their profiles directly. On the other hand, without specifying the purpose of clustering, we cannot decide which types of links to use for the clustering. Shall we use the friendship links to detect the social communities, or the user-like-book links to detect the reading groups, or a mix of them? Obviously, to solve such clustering tasks, we need to use both the incomplete attribute information as well as the link information of different types with the awareness of their importance weights. In our example, in order to discover a user's political interests, we need to learn which link types are more important for our purpose of clustering, among the relationships between her and blogs, books, and her friends.

Recently, some studies [42; 47; 59; 71; 75; 87] show that the combination of attribute and link information in a network can improve the clustering quality. However, none of them has addressed the two challenges simultaneously. Some of them rely on a complete attribute space and the clustering result is considered as a trade-off between attribute-based measures and link-based measures. Moreover, none of the current studies has examined the issue that different types of links have different importance in determining a clustering with a certain purpose.

Here we explore the interplay between different types of links and the specified attribute set in the clustering process and design a comprehensive and robust probabilistic clustering model for heterogeneous information networks.

6.2 THE RELATION STRENGTH-AWARE CLUSTERING PROBLEM DEFINITION

As defined before, a heterogeneous information network $G = (\mathcal{V}, \mathcal{E}, W)$ is modeled as a directed graph, where each node in the network corresponds to an object (or an event) in real life, and each link corresponds to a relationship between the linked objects. Associated with each link, there is a binary or positive value, denoting its input weight.

Attributes are associated with objects, such as the location of a user, the text description of a book, the text information of a blog, and so on. In this setting, we consider attributes across all different types of objects as a collection of attributes for the network, denoted as $\mathcal{X} = \{X_1, \ldots, X_T\}$, in which we are interested only in a subset for a certain clustering purpose. Each object $v \in \mathcal{V}$ contains a subset of the attributes, with **observations** denoted as $v[X] = \{x_{v,1}, x_{v,2}, \ldots, x_{v}, N_{X,v}\}$, where $N_{X,v}$ is the total number of observations of attribute X attached with object v. Note that some attributes can be shared by different types of objects, such as the text and the location attribute, while some other attributes are unique for a certain type of objects, such as the time length for a video clip. We use $\mathcal{V}_X$ to denote the object set that contains attribute X.

6.2.1 THE CLUSTERING PROBLEM

The goal of the clustering problem is to map every object in the network into a unified hidden space, that is, a soft clustering, according to the user-specified subset of attributes in the network, with the help of links from different types.

There are several new challenges for clustering objects in this new scenario. First, the attributes are usually **incomplete** for an object: the attributes specified by a user may be only partially or even not contained in an object type, and the values for these attributes could be missing even if the attribute type is contained in the object type. Moreover, the incompleteness of the data cannot be easily handled by interpolation: the observations for each attribute could be a set or a bag of values, and the neighbors for an object are from different types of objects, which may not be helpful for predicting the missing data. For example, it is impossible to get a user's blog via interpolating techniques. Therefore, none of the existing clustering algorithms that purely based on attribute space can solve the clustering problem in this scenario.

Second, with the awareness that links play a critical role to propagate the cluster information among objects, another challenge is that **different link types** have different semantic meanings and therefore have different strengths in the process of passing cluster information around. In other words, while it is clear that the existence of links between nodes is indicative of clustering similarity, it is also important to understand that *different link types may have a different level of importance in the clustering process*. In the example of clustering political interests illustrated in Figure 6.1, we expect a higher importance of the relation *user-like-book* than the relation *friendship* in deciding the cluster membership of a user. Thus, we need to design a clustering model which can learn the importance of these link types automatically. This will enhance the clustering quality because it marginalizes the impact of low quality types of neighbors of an object during the clustering process.

We present examples of clustering tasks in two concrete heterogeneous information networks in the following.

Example 6.1 (Bibliographic information network) A *bibliographic network* is a typical heterogeneous network, containing objects from three types of entities, namely *papers, publication venues* (conferences or journals), and *authors*. Each paper has different link types to its authors and publication venue. Each paper is associated with the text attribute as a bag of words. Each author and venue links to a set of papers, but contains no attributes (in our case). The application of a clustering process according to the text attribute in such a scenario can help detect research areas, and decide the research areas for authors, venues, and papers.

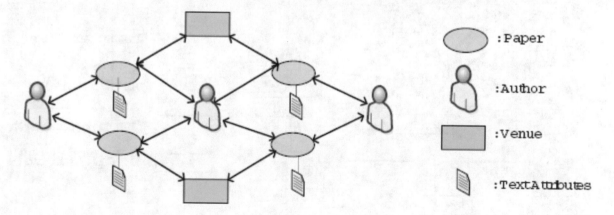

Figure 6.2: Illustration of bibliographic information network.

Note that we treat text as attributes of papers in this case instead of term entities as in previous chapters. Multiple types of objects and links in this network are illustrated in Figure 6.2. For objects of different types, their cluster memberships may need to be determined by different kinds of information: for authors and venues, the only available information is from the papers linked to them; for papers, both text attributes and links of different types are provided. Note that even for papers that are associated with text attributes, using link information can further help the clustering quality when the observations of the text data is very limited (e.g., using text merely from titles). Also, we may expect that the neighbors of an author type play a more important role in deciding a paper's cluster compared with the neighbor of a venue type. This needs to be automatically learned in terms of the underlying relation strengths.

Example 6.2 (Weather sensor network) Weather sensor networks typically contain different kinds of sensors for detecting different attributes, such as precipitation or temperature. Some sensors may have incorrect or no readings because of the inaccuracy or malfunctioning of the instruments. The links between sensors are generated according to their k nearest neighbors under geo-distances, in order to incorporate the importance of locality in weather patterns. The clustering of such sensors according to both precipitation and temperature attributes can be useful in determining regional weather patterns.

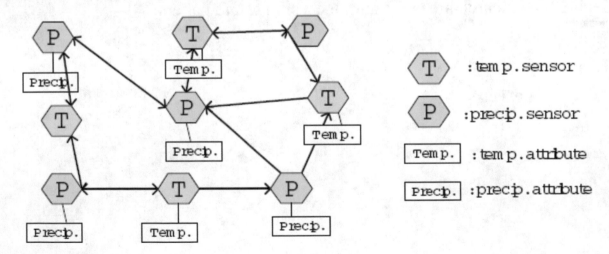

Figure 6.3: Illustration of weather sensor information network.

Figure 6.3 illustrates a weather sensor network containing two types of sensors: temperature and precipitation. A sensor may sometimes register none or multiple observations. Although it is desirable to use the complete observations on both temperature and precipitation to determine the weather pattern of a location, in reality a sensor object may contain only partial attribute (e.g., temperature values only for temperature sensors), and both of the attribute and link information are needed for correctly detecting the clusters. Still, which type of links plays a more important role needs to be determined in the clustering process.

Formally, given a network $G = (\mathcal{V}, \mathcal{E}, W)$, a specified subset of its associated attributes $X \in \mathcal{X}$, the attribute observations $\{v[X]\}$ for all objects, and the number of clusters K, our goal is:

1. to learn a soft clustering for all the objects $v \in \mathcal{V}$, denoted by a membership probability matrix, $\Theta_{|\mathcal{V}| \times K} = (\theta_v)_{v \in \mathcal{V}}$, where $\Theta(v, k)$ denotes the probability of object v in cluster k, $0 \leq \Theta(v, k) \leq 1$ and $\sum_{k=1}^{K} \Theta(v, k) = 1$, and θ_v is the K dimensional cluster membership vector for object v; and

2. to learn the strengths (importance weights) of different link types in determining the cluster memberships of the objects, $\gamma_{|\mathcal{R}| \times 1}$, where $\gamma(r)$ is a real number and stands for the importance weight for the link type $r \in \mathcal{R}$.

6.3 THE CLUSTERING FRAMEWORK

We propose a novel probabilistic clustering model in this section and introduce the algorithm that optimizes the model in Section 6.4.

6.3.1 MODEL OVERVIEW

Given a network G, with the observations of its links and the observations $\{v[X]\}$ for the specified attributes $X \in \mathcal{X}$, a good clustering configuration Θ, which can be viewed as hidden cluster information for objects, should satisfy two properties.

1. Given the clustering configuration, the observed attributes should be generated with a high probability. Especially, we model each attribute for each object as a separate mixture model, with each component representing a cluster.

2. The clustering configuration should be highly consistent with the network structure. In other words, linked objects should have similar cluster membership probabilities, and larger strength of a link type requires more similarity between the linked objects of this type.

Overall, we can define the likelihood of the observations of all the attributes $X \in \mathcal{X}$ as well as the hidden continuous cluster configuration Θ, given the underneath network G, the relation strength vector γ, and the cluster component parameter β, which can be decomposed into two parts, the generative probability of the observed attributes given Θ and the probability of Θ given the network structure:

$$p(\{\{v[X]\}_{v \in \mathcal{V}_X}\}_{X \in \mathcal{X}}, \Theta | G, \gamma, \beta) = \prod_{X \in \mathcal{X}} p(\{v[X]\}_{v \in \mathcal{V}_X} | \Theta, \beta) p(\Theta | G, \gamma) . \qquad (6.1)$$

From a generative point of view, this model explains how observations for attributes associated with objects are generated: first, a hidden layer of variables Θ is generated according to the probability $p(\Theta | G, \gamma)$, given the network structure G and the strength vector γ; second, the observed values of attributes associated with each object are generated according to mixture models, given the cluster membership of the object, as well as the cluster component parameter β, with the probability $\prod_{X \in \mathcal{X}} p(\{v[X]\}_{v \in \mathcal{V}_X} | \Theta, \beta)$.

The goal is then to find the best parameters γ and β, as well as the best clustering configuration Θ that maximize the likelihood. The detailed modeling of the two parts is introduced in the following.

6.3.2 MODELING ATTRIBUTE GENERATION

Given a configuration Θ for the network G, namely, the membership probability vector θ_v for each object v, the attribute observations for each object v are conditionally independent with observations from other objects. Each attribute X associated with each object v is then assumed following the same family of mixture models that share the same cluster components, with the component mixing proportion as the cluster membership vector θ_v. For simplicity, we first assume that only one attribute X is specified for the clustering purpose and then briefly discuss a straightforward extension to the multi-attribute case.

Single Attribute

Let X be the only attribute we are interested in the network, and let $v[X]$ be the observed values for object v, which may contain multiple observations. It is natural to consider that the attribute observation $v[X]$ for each object v is generated from a mixture model, where each component is a probabilistic model that stands for a cluster, with the parameters to be learned, and the component weights denoted by θ_v. Formally, the probability of all the observations $\{v[X]\}_{v \in \mathcal{V}_X}$ given the network configuration Θ is modeled as:

$$P(\{v[X]\}_{v \in \mathcal{V}_X} | \Theta, \beta) = \prod_{v \in \mathcal{V}_X} \prod_{x \in v[X]} \sum_{k=1}^{K} \theta_{v,k} p(x|\beta_k), \tag{6.2}$$

where K is the number of clusters and β_k is the parameter for component k. In this chapter, we consider two types of attributes, one corresponding to text attributes with categorical distributions, and the other numerical attributes with Gaussian distributions.

1. **Text attribute with categorical distribution**. In this case, objects in the network contain text attributes in the form of a term list, from the vocabulary $l = 1$ to m. Each cluster k has a different term distribution following a categorical distribution, with the parameter $\beta_k = (\beta_{k,1}, \ldots, \beta_{k,m})$, where $\beta_{k,l}$ is the probability of term l appearing in cluster k, that is, $X|k \sim discrete(\beta_{k,1}, \ldots, \beta_{k,m})$. Following the frequently used topic modeling method PLSA [25], each term in the term list for an object v is generated from the mixture model, with each component as a categorical distribution over terms described by β_k, and the component coefficient is

θ_v. Formally, the probability of observing all the current attribute values is

$$p(\{v[X]\}_{v \in \mathcal{V}_X} | \Theta, \beta) = \prod_{v \in \mathcal{V}_X} \prod_{l=1}^{m} \left(\sum_{k=1}^{K} \theta_{v,k} \beta_{k,l} \right)^{c_{v,l}}, \tag{6.3}$$

where $c_{v,l}$ denotes the count of term l that object v contains.

2. **Numerical attribute with Gaussian distribution.** In this case, objects in the network contain numerical observations in the form of a value list, from the domain $\mathbb{R}$. The kth cluster is a Gaussian distribution with parameters $\beta_k = (\mu_k, \sigma_k^2)$, that is, $X|k \sim \mathcal{N}(\mu_k, \sigma_k^2)$, where μ_k and σ_k are mean and standard deviation of normal distribution for component k. Each observation in the observation list for an object v is generated from the Gaussian mixture model, where each component is a Gaussian distribution with parameters μ_k, σ_k^2, and the component coefficient is θ_v. The probability density for all the observations for all objects is then:

$$p(\{v[X]\}_{v \in \mathcal{V}_X} | \Theta, \beta) = \prod_{v \in \mathcal{V}_X} \prod_{x \in v[X]} \sum_{k=1}^{K} \theta_{v,k} \frac{1}{\sqrt{2\pi\sigma_k^2}} e^{-\frac{(x-\mu_k)^2}{2\sigma_k^2}}. \tag{6.4}$$

Multiple Attributes

As in the weather sensor network example, we are interested in multiple attributes, namely temperature and precipitation. Generally, if multiple attributes in the network are specified by users, say $X_1, \ldots, X_T$, the probability density of observed attribute values $\{v[X_1]\}, \ldots, \{v[X_T]\}$ for a given clustering configuration Θ is as follows, by assuming the independence among these attributes:

$$p(\{v[X_1]\}_{v \in \mathcal{V}_{X_1}}, \ldots, \{v[X_T]\}_{v \in \mathcal{V}_{X_T}} | \Theta, \beta_1, \ldots, \beta_T) = \prod_{t=1}^{T} p(\{v[X_t]\}_{v \in \mathcal{V}_{X_t}} | \Theta, \beta_t). \tag{6.5}$$

6.3.3 MODELING STRUCTURAL CONSISTENCY

From the view of links, the more similar the two objects are in terms of cluster memberships, the more likely they are connected by a link. In order to quantitatively measure the consistency of a clustering result Θ with the network structure G, we define a novel probability density function for observing Θ.

We assume that linked objects are more likely to be in the same cluster, if the link type is of importance in determining the clustering process. That is, for two linked objects, v_i and v_j, their membership probability vectors θ_i and θ_j should be similar. Within the same type of links, the higher link weight ($w(e)$), the more similar θ_i and θ_j should be. Further, a certain link type may be of greater importance, and will influence the similarity to a greater extent. The consistency of a configuration Θ with the network G, is evaluated with the use of a composite analysis with respect to all the links in the network in the form of a probability density value. A more consistent configuration of Θ will yield a higher probability density value. In the following, we first introduce how the consistency of two cluster membership vectors is defined with respect to a single link, and then show how this analysis can be applied over all links in order to create a probability density value as a function of Θ.

For a link $e = \langle v_i, v_j \rangle \in \mathcal{E}$, with type $r = \phi(e) \in \mathcal{R}$, we denote the *importance of the link type to the clustering process* by a real number $\gamma(r)$. This is different from the weight of the link $w(e)$, which is specified in the network as input, whereas the value of $\gamma(r)$ is defined on link types and needs to be learned. We denote the consistency function of two cluster membership vectors, θ_i and θ_j, with link e under strength weights for each link type γ by a *feature function* $f(\theta_i, \theta_j, e, \gamma)$. Higher values of this function imply greater consistency with the clustering results. In the following, we list several desiderata for a good feature function.

1. The value of the feature function f should increase with greater similarity of θ_i and θ_j.

2. The value of the feature function f should decrease with greater importance of the link e, either in terms of its specified weight $w(e)$, or the learned importance $\gamma(r)$ for its link type. In other words, for the larger

strength of a particular link type, two linked nodes are required to be more similar in order to claim the same level of consistency.

3. The feature function should not be symmetric between its first two arguments θ_i and θ_j, because the impact from node v_i to node v_j could be different from that of v_j to v_i.

The last criterion may need some further explanation. For example, in a citation network, a paper i may cite paper j, because i feels that j is relevant to itself, while the reverse may not be necessarily true. In the experimental section, we will show that asymmetric feature functions produce higher accuracy in link prediction.

We then propose a cross entropy-based feature function, which satisfies all of the desiderata listed above. For a link $e = \langle v_i, v_j \rangle \in \mathcal{E}$, with relation type $r = \phi(e) \in \mathcal{R}$, the feature function $f(\theta_i, \theta_j, e, \gamma)$ is defined as:

$$f(\theta_i, \theta_j, e, \gamma) = -\gamma(r)w(e)H(\theta_j, \theta_i) = \gamma(r)w(e)\sum_{k=1}^{K} \theta_{j,k} \log \theta_{i,k} . \qquad (6.6)$$

where, $H(\theta_j, \theta_i) = -\sum_{k=1}^{K} \theta_{j,k} \log \theta_{i,k}$, is the cross entropy from θ_j to θ_i, which evaluates the deviation of v_j from v_i, in terms of the average coding bits needed if using coding schema based on the distribution of θ_i. For a fixed value of $\gamma(r)$, the value of $H(\theta_j, \theta_i)$ is minimal and (therefore) f is maximal, when the two vectors are identical. It is also evident from Equation (6.6) that the value of f decreases with increasing learned link type strength $\gamma(r)$ or input link weight $w(e)$. We require $\gamma \geq 0$, in the sense that we do not consider links that connect dissimilar objects. The value of f so defined is a non-positive function, with larger value indicating a higher consistency of the link.

Other distance functions such as KL-divergence could replace the cross entropy in the feature function. However, as cross entropy favors distributions that concentrate on one cluster ($H(\theta_j, \theta_i)$ achieves the lowest distance, when $\theta_j = \theta_i$ and $\theta_{i,k} = 1$ for some cluster k), which agrees with our clustering purpose, we pick it over KL-divergence.

We then propose a log-linear model to model the probability of Θ given the link type weights γ, where the probability of one configuration Θ is defined as the exponential of the summation of feature functions of all the links in G:

$$p(\Theta|G, \gamma) = \frac{1}{Z(\gamma)} \exp\{ \sum_{e=\langle v_i, v_j\rangle \in \mathcal{E}} f(\theta_i, \theta_j, e, \gamma)\} \,, \tag{6.7}$$

where γ is the strength weight vector for all link types, $f(\theta_i, \theta_j, e, \gamma)$ is the feature function defined on links of different types, and $Z(\gamma)$ is the partition function that makes the distribution function sum up to 1: $Z(\gamma) = \int_\Theta \exp\{\sum_{e=\langle v_i,v_j\rangle \in \mathcal{E}} f(\theta_i, \theta_j, e, \gamma)\}d\Theta$. The partition function $Z(\gamma)$ is an integral over the space of all the configurations Θ, and it is a function of γ.

6.3.4 THE UNIFIED MODEL

The overall goal of the network clustering problem is to determine the best clustering results Θ, the link type strengths γ, and the cluster component parameters β that maximize the generative probability of attribute observations and the consistency with the network structure, described by the likelihood function in Equation (6.1).

Further, we add a Gaussian prior to γ as a regularization to avoid overfitting, with the mean as 0, and the covariance matrix as $\sigma^2 I$, where σ is the standard deviation of each element in γ, and I is the identity matrix. We set $\sigma = 0.1$ in our experiments, and more complex strategy can be used to select σ according to labeled clustering results, which will not be discussed here. The new objective function is then:

$$g(\Theta, \beta, \gamma) = \log \sum_{X \in \mathcal{X}} p(\{v[X]\}_{v \in V_X}|\Theta, \beta) + \log p(\Theta|G, \gamma) - \frac{\|\gamma\|^2}{2\sigma^2}. \tag{6.8}$$

In addition, we have the constraints that $\gamma \geq 0$, and some constraints for β that are dependent on the attribute distribution type. Also, $p(\{v[X]\}_{v \in V_X}|\Theta, \beta)$ and $p(\Theta|G, \gamma)$ need to be replaced by the specific formulas proposed above for concrete derivations.

6.4 THE CLUSTERING ALGORITHM

This section presents a clustering algorithm that computes the proposed probabilistic clustering model. Intuitively, we begin with the assumption that all the types of links play an equally important role in the clustering process, then update the strength for each type according to the average consistency of links of that type with the current clustering results, and finally achieve a good clustering as well as a reasonable strength vector for link types. It is an iterative algorithm containing two steps in that clustering results and strengths of link types mutually enhance each other, which maximizes the objective function of Equation (6.8) alternatively.

In the first step, we fix the link type weights γ to the best value γ^*, determined in the last iteration, then the problem becomes that of determining the best clustering results Θ and the attribute parameters β for each cluster component. We refer to this step as the *cluster optimization step*: $[\Theta^*, \beta^*] = \arg\max_{\Theta,\beta} g(\Theta, \beta, \gamma^*)$.

In the second step, we fix the clustering configuration parameters $\Theta = \Theta^*$ and $\beta = \beta^*$, corresponding to the values determined in the last step, and use it to determine the best value of γ, which is consistent with current clustering results. We refer to this step as the *link type strength learning step*: $\gamma^* = \arg\max_{\gamma \geq 0} g(\Theta^*, \beta^*, \gamma)$.

The two steps are repeated until convergence is achieved.

6.4.1 CLUSTER OPTIMIZATION

In the cluster optimization step, each object has the link information from different types of neighbors, where the strength of each type of link is given, as well as the possible attribute observations. The goal is to utilize both link and attribute information to get the best clustering result for all the objects. Since γ is fixed in this step, the partition function and regularizer term become constants, and can be discarded for optimization purposes. Therefore, we can construct a simplified objective function $g_1(\cdot, \cdot)$, which depends only on Θ and β:

$$g_1(\Theta, \beta) = \sum_{e=\langle v_i, v_j \rangle} f(\theta_i, \theta_j, e, \gamma) + \sum_{v \in \mathcal{V}_X} \sum_{x \in v[X]} \log \sum_{k=1}^{K} \theta_{v,k} P(x|\beta_k) . \tag{6.9}$$

We derived an EM-based algorithm [8; 17] to solve Equation (6.9). In the E-step, the probability of each observation x for each object v and each attribute X belonging to each cluster, usually called the hidden cluster label of the observation, $z_{v, x}$, is derived according to the current parameters Θ and β. In the M-step, the parameters Θ and β are updated according to the new membership for all the observations in the E-step. The iterative formulas for single text attribute and single Gaussian attribute are provided below.

1. **Single categorical text attribute**. Let $z_{v, l}$ denote the hidden cluster label for the lth term in the vocabulary for object v, Θ^{t-1} be the value of Θ at iteration $t - 1$, and β^{t-1} be the value of β at iteration $t - 1$. $\mathbf{1}_{\{v \in \mathcal{V}_X\}}$ is the indicator function, which is 1 if v contains this attribute, and 0 otherwise. Then, we have:

$$p(z_{v,l}^t = k|\Theta^{t-1}, \beta^{t-1}) \propto \theta_{v,k}^{t-1} \beta_{k,l}^{t-1}$$

$$\theta_{v,k}^t \propto \sum_{e=\langle v,u \rangle} \gamma(\phi(e)) w(e) \theta_{u,k}^{t-1} + \mathbf{1}_{\{v \in \mathcal{V}_X\}} \sum_{l=1}^{m} c_{v,l} p(z_{v,l}^t = k|\Theta^{t-1}, \beta^{t-1}) \tag{6.10}$$

$$\beta_{k,l}^t \propto \sum_{v \in \mathcal{V}_X} c_{v,l} p(z_{v,l}^t = k|\Theta^{t-1}, \beta^{t-1}) .$$

2. **Single Gaussian numerical attribute**. Let $z_{v, x}$ denote the hidden cluster label for the observation x for object v, Θ^t be the value of Θ at iteration t, and μ_k^t and σ_k^t be the values of mean and standard deviation for kth cluster at iteration t. $\mathbf{1}_{\{v \in \mathcal{V}_X\}}$ is the indicator function, which is 1 if v contains this attribute, and 0 otherwise. Then, we have:

$$p(z_{v,x}^t = k | \Theta^{t-1}, \beta^{t-1}) \propto \theta_{v,k}^{t-1} \frac{1}{\sqrt{2\pi(\sigma_k^{t-1})^2}} e^{-\frac{(x-\mu_k^{t-1})^2}{2(\sigma_k^{t-1})^2}}$$

$$\theta_{v,k}^t \propto \sum_{e=(v,u)} \gamma(\phi(e)) w(e) \theta_{u,k}^{t-1} + 1_{\{v \in V_X\}} \sum_{x \in v[X]} p(z_{v,x}^t = k | \Theta^{t-1}, \beta^{t-1}) \quad (6.11)$$

$$\mu_k^t = \frac{\sum_{v \in V_X} \sum_{x \in v[X]} x p(z_{v,x}^t = k | \Theta^{t-1}, \beta^{t-1})}{\sum_{v \in V_X} \sum_{x \in v[X]} p(z_{v,x}^t = k | \Theta^{t-1}, \beta^{t-1})}$$

$$(\sigma_k^2)^t = \frac{\sum_{v \in V_X} \sum_{x \in v[X]} (x - \mu_k^t)^2 p(z_{v,x}^t = k | \Theta^{t-1}, \beta^{t-1})}{\sum_{v \in V_X} \sum_{x \in v[X]} p(z_{v,x}^t = k | \Theta^{t-1}, \beta^{t-1})}.$$

For networks with multiple attributes, the formulae can be derived similarly. The readers can find the formulae for the case of two Gaussian numerical attributes in [61].

From the update rules, we can see that the value of the membership probability for an object is dependent on its neighbors' memberships, the strength of the link types, the weight of the links, and the attribute associated with it (if any). When an object contains no attributes in the specified set, or contains no observations for the specified attributes, the cluster membership is totally determined by its linked objects, which is a weighted average of their cluster memberships and the weight is determined by both the weight of the link and the weight of the link type. When an object contains some observations of the specified attributes, its cluster membership is determined by both its neighbors and these observations for each possible attribute.

6.4.2 LINK TYPE STRENGTH LEARNING

The link type strength learning step is to find the best strength weight for each type of links that makes the current clustering result to be generated with the highest probability. By doing so, the low-quality link types that connect objects not so similar will be punished and assigned with low strength weights; while the high quality link types will be assigned with high strength weights.

Since the values of Θ and β are fixed in this step, the only relevant parts of the objective function (for optimization purposes) are those which depend on γ. These are the structural consistency modeling part and the regularizer over γ. Therefore, we can construct the following simplified objective function $g_2(\cdot)$ as a function of γ:

$$g_2(\gamma) = \sum_{e=\{v_i,v_j\}} f(\theta_i, \theta_j, e, \gamma) - \log Z(\gamma) - \frac{||\gamma||^2}{2\sigma^2}. \tag{6.12}$$

In addition, we have the linear constraints as $\gamma \geq 0$.

However, g_2 is difficult to be optimized directly, since the partition function $Z(\gamma)$ is an integral over the entire space of valid values of Θ, which is intractable. Instead, we construct an alternate approximate objective function g_2', which factorizes $\log p(\Theta|G)$ as the sum of $\log p(\theta_i|\theta_{-i}, G)$, namely the pseudo-log-likelihood, where $p(\theta_i|\theta_{-i}, G)$ is the conditional probability of θ_i given the remaining objects' clustering configurations, which turns out to be dependent only on its neighbors. The intuition of using pseudo-log-likelihood to approximate the real log-likelihood is that, if the probability of generating the clustering configuration for each object conditional on its neighbors is high, the probability of generating the whole clustering configuration should also be high. In other words, if the local patches of a network are very consistent with the clustering results, the consistency over the whole network should also be high.

In particular, we choose each local patch of the network as an object and all its out-link neighbors. In this case, every link is considered exactly once, and the newly designed objective function $g_2'(\cdot)$ is as follows:

$$g_2'(\gamma) = \sum_{i=1}^{|V|} \left(\sum_{e=\{v_i,v_j\}} f(\theta_i, \theta_j, e, \gamma) - \log Z_i(\gamma) \right) - \frac{||\gamma||^2}{2\sigma^2}, \tag{6.13}$$

where $\log Z_i(\gamma) = \log \int_{\theta_i} e^{\sum_{e=\{v_i,v_j\}} f(\theta_i,\theta_j,e,\gamma)} d\theta_i$, the local partition function for object v_i, with the linear constraints $\gamma \geq 0$.

As the joint distribution of Θ as well as the conditional distribution of θ_i given its out-link neighbors are both belonging to exponential families, both g_2 and g_2' are concave functions of γ. Therefore, the maximum value is either achieved at the global maximum point or at the boundary of constraints. The Newton-Raphson method is used to solve the optimization problem. It needs to calculate the first and second derivative of $g'_2(\gamma)$ with respect to γ, which is non-trivial in our case. We discuss the computation of these below.

By re-examining $p(\theta_i|\{\theta_j\}_{\forall e=\langle v_i,v_j\rangle}, G)$, the conditional probability for each object i given its out-link neighbors, we have:

$$p(\theta_i|\{\theta_j\}_{\forall e=\langle v_i,v_j\rangle}, G) \propto \prod_{k=1}^{K} \theta_{ik}^{\sum_{e=\langle v_i,v_j\rangle} \gamma(\phi(e))w(e)\theta_{j,k}}. \tag{6.14}$$

It is easy to see that $p(\theta_i|\{\theta_j\}_{\forall e=\langle v_i,v_j\rangle}, G)$ is a Dirichlet distribution with parameters $\alpha_{ik} = \sum_{e=\langle v_i,v_j\rangle} \gamma(\phi(e))w(e)\theta_{j,k} + 1$, for $k = 1$ to K. Therefore, the local partition function for each object i, $Z_i(\gamma)$, should be the constant $B(\alpha_i)$ as in Dirichlet distribution, where $\alpha_i = (\alpha_{i1}, \ldots, \alpha_{iK})$ and $B(\alpha_i) = \frac{\prod_{k=1}^{K} \Gamma(\alpha_{ik})}{\Gamma(\sum_{k=1}^{K} \alpha_{ik})}$. Then the first and second derivatives ($\nabla g'_2$ and Hg'_2) can be calculated now as each Z_i is a function of Gamma functions. Then, we can use the Newton-Raphson method to determine the value of γ that maximizes g'_2 with the following iterative steps:

1. $\gamma^{t+1} = \gamma^t - [Hg'_2(\gamma^t)]^{-1}\nabla g'_2(\gamma^t)$;

2. $\forall r \in \mathcal{R}$, if $\gamma(r)^{t+1} < 0$, set $\gamma(r)^{t+1} = 0$.

6.4.3 PUTTING TOGETHER: THE GENCLUS ALGORITHM

We integrate the two steps discussed above to construct a **Gen**eral Heterogeneous Network **Clus**tering algorithm, *GenClus*. The algorithm includes an outer iteration that updates Θ and γ alternatively, and two inner iterations that optimize Θ using the EM algorithm and optimize γ using the Newton-Raphson method, respectively. For the initialization of γ in the outer iteration, we initialize it as an all-1 vector. This means that all the link types in the network are initially considered equally important. For the initialization of Θ' in the inner iteration for optimizing Θ, we can either (1) assign Θ'^0 with random assignments, or (2) start with several random seeds, run the EM algorithm for a few steps for each random seed, and choose the one with the highest value of the objective function g_1 as the real starting point. The latter approach will produce more stable results.

The time complexity for the EM algorithm in the first step is $O(t_1(Kd_1|\mathcal{V}| + K|\mathcal{E}|))$, where t_1 is the number of iterations, d_1 is the average number of observations for each object, K is the number of clusters, $|\mathcal{V}|$ is the number of objects, and $|\alpha|$ is the number of links in the network, which is linear to $|\mathcal{V}|$ for sparse networks. The time complexity of the algorithm in the step of maximizing γ is dependent on the time for calculating the first derivative and Hessian matrix of $g_2'(\gamma)$, and the matrix inversion involved Newton-Raphson algorithm. This is $O(K|\mathcal{E}|+ t_2|\mathcal{R}|^{2.376}))$, where K and $|\mathcal{E}|$ are with the same meaning as before, t_2 is the number of iterations, and $|\mathcal{R}|$ is the number of relations in the network. In all, the overall time complexity is $O(t(t_1(Kd_1|\mathcal{V}| + K|\mathcal{E}|) + t_2|\mathcal{R}|^{2.376}))$, where t is the number of outer iterations. In other words, for each outer iteration, the time complexity is approximately linear to the number of objects in the network when the network is sparse. Therefore, the *GenClus* algorithm is quite scalable.

6.5 EXPERIMENTAL RESULTS

In this section, we examine the effectiveness of *GenClus* on several real and synthetic datasets.

6.5.1 DATASETS

Two real networks and one synthetic network are used in this study. We extracted two networks from the *DBLP "four-area" dataset* [21; 69], by using different subsets of entities and the links between them to represent the underlying network structures. This dataset was extracted from 20 major conferences from the 4 areas corresponding to database, data mining, information retrieval, and artificial intelligence. Besides the real networks, we also generated a synthetic weather sensor network. We describe these networks below.

(a) **DBLP four-area A-V network.** This network contains two types of objects, authors (A) and venues (V); and three types of links depending upon publication behavior, namely *publish_in(A, V)* (short for $\langle A, V \rangle$), *published_by(V, A)* (short for $\langle V, A \rangle$), and *coauthor(A, A)* (short for $\langle A, A \rangle$). The links are associated with a weight corresponding to the number of papers that an author has published in a venue, a venue is contributed by an author, and the two authors have co-authored, respectively. The author nodes and venue nodes contain text corresponding to the text from the titles of all the papers they have ever written or published.

(b) **DBLP four-area A-V-P network.** This network contains objects corresponding to authors (A), venues (V), and papers (P); and four types of links depending upon the publication behavior, namely *write(A, P)* (short for $\langle A, P \rangle$), *written_by(P, A)* (short for $\langle P, A \rangle$), *publish(V, P)* (short for $\langle V, P \rangle$), and *published_by(P, V)* (short for $\langle P, V \rangle$). In this case, the links have binary weights, corresponding to presence or absence of the link. Only papers contain text attributes that are from their titles.

(c) **Weather sensor network.** This network is synthetically generated, containing two types of objects: temperature (T) and precipitation (P) sensors, and four link types between any two types of sensors denoting the kNN relationship: $\langle T, T \rangle$, $\langle T, P \rangle$, $\langle P, T \rangle$, and $\langle P, P \rangle$. The links are binary weighted according to their *k*-nearest neighbors. The attributes associated with a sensor correspond to either temperature or precipitation, depend on the type of the sensor. We use the weather network generator to generate two sets of synthetic climate sensor networks, each containing four clusters, and each sensor is linked to five nearest neighbors for each type (ten in total). In each setting, we vary the number of sensors, by fixing the number of temperature sensors at 1000, and precipitation sensors as 250, 500, and 1000. For each setting, the number of observations for each object may be 1, 5, or 20. In all, for each weather pattern setting, we have 9 networks with different configurations.

6.5.2 EFFECTIVENESS STUDY

We use two measures for our effectiveness study. First, the labels associated with the nodes in the datasets provide a natural guidance in examining the coherence of the clusters. We use *Normalized Mutual Information (NMI)* [60] to compare our clustering result with the ground truth. Second, we use link prediction accuracy to test the clustering accuracy. The similarity between two objects can be calculated by similarity function defined on their two membership vectors, such as using cosine similarity function. Clearly, a better clustering quality will lead to better computation of similarity (and therefore the accuracy of link prediction). For a certain type of relation $\langle A, B \rangle$, we calculate all the similarity scores between each $v_A \in A$ and all the objects $v_B \in B$, and compare the similarity-based ranked list with the true ranked list determined by the link weights between them. We use the measure *Mean Average Precision (MAP)* [81] to compare the two ranked links.

Clustering Accuracy Test We choose clustering methods that can deal with both the links and attributes as our baselines. None of these baselines is capable of leveraging different link types of different impacts to the clustering process. Therefore, we set each link type strength as 1 for these baselines. Second, we choose different baselines for clustering networks with text attributes and for clustering networks with numerical attributes, since there are no unified clustering methods (other than our presented *GenClus*) that can address both situations in the same framework.

For the *DBLP four-area A-V network* and the *DBLP four-area A-V-P network* that are with text attributes, we use *NetPLSA* [47] and *iTopicModel* [64] as baselines, which aim at improving topic qualities by using link information in homogeneous networks. We compare *GenClus* with the baselines by assuming homogeneity of links for the latter algorithms. The mean and standard deviation of NMI of the 20 running results are shown for the *DBLP A-V network* and the *DBLP A-V-P network* in Figures 6.4 and 6.5, respectively. From the results, we can see that the *GenClus* algorithm is much more effective than both the *iTopicModel* and the *NetPLSA* methods in both networks. This is because of the ability of the former algorithms to learn and leverage the strengths of different link types in the clustering process. Furthermore, the standard deviation of NMI over different runs is much lower for *GenClus*, which suggests that the algorithm is more robust to the initial settings with the learned strength weights for different link types.

The *A-V network* is the easiest case among the three networks, since it only contains one type of attribute (the text attribute), and all object types contain this attribute, namely the attribute is complete for every object. The *A-V-P network* is a more difficult case than the previous one, because not every type of objects contain the text attributes. This requires the clustering algorithm to be more robust to deal with objects with no attributes at all. From the results, we can see that *GenClus* is more robust than *NetPLSA* algorithm, which outputs almost random predictions for authors for the *A-V-P network*. Although the homogenous methodology of the *iTopicModel* algorithm performs better for objects of type V for *A-V network* (see Figure 6.5), *GenClus* still has an overall better performance. This is because our objective function is over all the objects rather than a particular type.

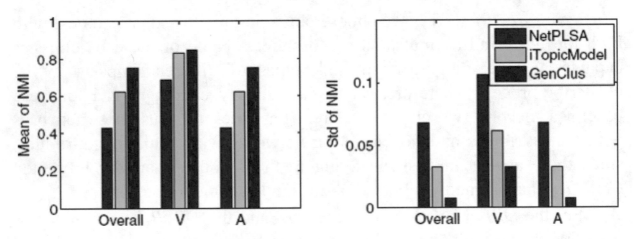

Figure 6.4: Clustering accuracy comparisons for the *A-V network*.

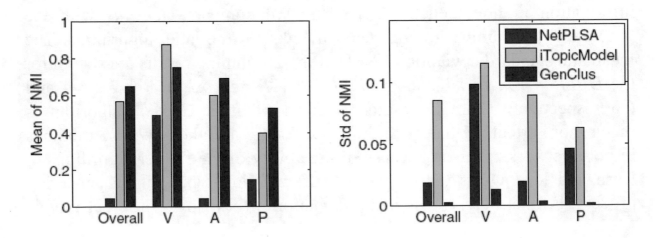

Figure 6.5: Clustering accuracy comparisons for the *A-V-P network*.

We also examined the actual clusters obtained by the algorithm on *DBLP A-V network*, and list corresponding cluster membership for several venues and authors in Table 6.1, where the research area names are given afterwards according the clustering results. We can see that the clustering results for the *GenClus* algorithm are consistent with human intuition.

The synthetic weather sensor network is the most difficult case among the three networks, as it has two types of attributes corresponding to different types of sensors. Furthermore, all sensor nodes contain incomplete observations of the attributes. Existing algorithms cannot address these issues well. We compare the clustering results of *GenClus* with two baselines, by comparing the cluster labels with maximum probabilities with the ground truth. In this case, we choose the initial seed for *GenClus* as one of the tentative running results with the highest objective function, and the number of iterations is set to five. The first baseline is the *k*-means algorithm, and the second one is a spectral clustering method that combines the network structure and attribute similarity as a new similarity matrix. We use the framework given in [59], which utilizes modularity objective function in the network part, but we replace the cosine similarity by Euclidean distance in the attribute part as in [80] for better clustering results. As both methods cannot handle the problem of incomplete attributes, we use interpolation to make each sensor have a regular two-dimensional attribute, by using the mean of all the observations of its neighbors and itself. For the spectral clustering-based framework, we centralize the data by extracting the mean and then normalize them by the standard deviation, in order to make the attribute part comparable with the modularity part in the objective function. Both parts are set to have equal weights.

Table 6.1: Case studies of cluster membership results

Object	DB	DM	IR	AI
SIGMOD	0.8577	0.0492	0.0482	0.0449
KDD	0.0786	0.6976	0.1212	0.1026
CIKM	0.2831	0.1370	0.4827	0.0971
Jennifer Widom	0.7396	0.0830	0.1061	0.0713
Jim Gray	0.8359	0.0656	0.0536	0.0449
Christos Faloutsos	0.4268	0.3055	0.1380	0.1296

The results are summarized in Figures 6.6 and 6.7. It is evident that the *GenClus* algorithm exhibits superior performance to the two baselines in most of the datasets (17 out of 18 cases). Furthermore, *GenClus* can produce more stable clustering results compared with *k*-means, which is very sensitive to the number of observations for each object, especially for Setting 2. *GenClus* is also highly adaptive in that there is no need of any weight specification for combining the network and attribute-contributions to the clustering process. This results in greater stability of *GenClus*. Another major advantage of *GenClus* (which is not immediately evident from the presented results) is that we can directly utilize every observation instead of the mean, while the baselines can only use a biased mean value because of the interpolation process.

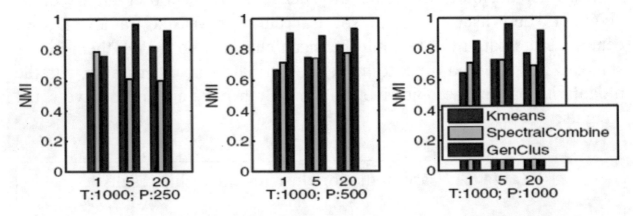

Figure 6.6: Clustering accuracy comparisons for weather sensor network Setting 1.

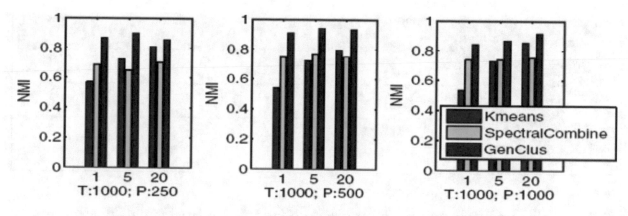

Figure 6.7: Clustering accuracy comparisons for weather sensor network Setting 2.

Link Prediction Accuracy Test Next, the link prediction accuracy measured by MAP is compared between *GenClus* and the baselines. For the *A-V network*, we select the link type $\langle A, V \rangle$ for the prediction task, namely we want to predict which venues an author is likely to go. For the *A-V-P network*, we select the link type $\langle P, V \rangle$ for the prediction task, namely we want to predict which venue a paper is published in. As the prediction is based on the similarity between the two objects, say query object v_i with clustering membership θ_i and candidate object v_j with clustering membership θ_j, three similarity functions are used here: (1) cosine similarity denoted as $\cos(\theta_i, \theta_j)$; (2) the negative of Euclidean distance denoted as $-\|\theta_i - \theta_j\|$; and (3) the negative of cross entropy denoted as $-H(\theta_j, \theta_i)$. The results are summarized in Tables 6.2 and 6.3.

Table 6.2: Prediction accuracy for A-V relation in *A-V network*

	NetPLSA	iTopicModel	GenClus
$\cos(\theta_i, \theta_j)$	0.4351	0.5117	**0.7627**
$-\|\theta_i - \theta_j\|$	0.4312	0.5010	**0.7539**
$-H(\theta_j, \theta_i)$	0.4323	0.5088	**0.7753**

Table 6.3: Prediction accuracy for P-V relation in *A-V-P network*

	NetPLSA	iTopicModel	GenClus
$\cos(\theta_i, \theta_j)$	0.2762	0.4609	**0.5170**
$-\|\theta_i - \theta_j\|$	0.2759	0.4600	**0.5142**
$-H(\theta_j, \theta_i)$	0.2760	0.4683	**0.5183**

For the weather sensor network, we select the link type $\langle T, P \rangle$, namely we want to predict the P-typed neighbors for T-typed sensors. We test the link prediction in the network with configuration as in Setting 1, with $\#T = 1000$ and $\#P = 250$. We only output the link prediction results for *GenClus* algorithm, since the other two baselines can only output hard clusters (exact cluster memberships rather than probabilities). The results are shown in Table 6.4.

Table 6.4: Prediction accuracy for $\langle T, P \rangle$ in weather sensor network

	$\cos(\theta_i, \theta_j)$	$-\|\theta_i - \theta_j\|$	$-H(\theta_j, \theta_i)$
MAP	0.7285	0.7690	**0.8073**

From the results, it is evident that the *GenClus* algorithm has the best link prediction accuracy in terms of different similarity functions. Also, the results show that the asymmetric function $-H(\theta_j, \theta_i)$ provides the best link prediction accuracy, especially for better clustering results such as those obtained by *GenClus* and in the weather sensor network where the out-link neighbors are different from the in-link neighbors.

Analysis of Link Type Strength Since the process of learning the semantic importance of relations is important in a heterogeneous clustering approach, we present the learned relation strengths in Figure 6.8 for the two *DBLP four-area networks*. From the figure, it is evident that in the *A-V network*, the link type $\langle A, V \rangle$ has greater importance to the clustering process than the link type $\langle A, A \rangle$, and thus is more important in deciding an author's membership. This is because the spectrum of co-authors is often broad, whereas authors' publication frequency in each venue can be a more reliable predictor of clustering behavior. For the *A-V-P network*, we can see that the link type $\langle P, V \rangle$ has the weight 3.13, whereas the link type $\langle P, A \rangle$ has a much higher weight 13.30. This suggests that the latter link type is more reliable in deciding the cluster for papers, since the venues usually have a broader research track than the authors. For example, it is difficult to judge the cluster for a paper if we only know that it is published in the CIKM venue. The ability of our algorithm to learn such important characteristics of different link types is one of the reasons that it is superior to other competing methods.

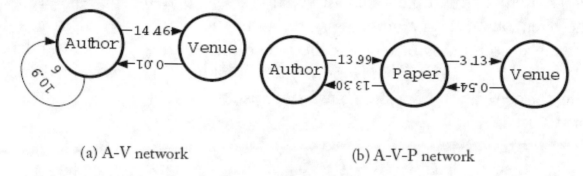

(a) A-V network   (b) A-V-P network

Figure 6.8: Strengths for link types in two *DBLP four-area networks*.

CHAPTER 7

User-Guided Clustering via Meta-Path Selection

In this chapter, we study another relation strength-aware mining problem: user-guided clustering of a certain type of objects, based on their involvement of multiple types of relations, encoded by meta-paths, in a heterogeneous information network. In an application, a user often has the best say on the kinds of clusters she would like to get, and such guidance will lead to the selection of appropriate combination of weighted meta-paths for generation of desired clustering results.

7.1 OVERVIEW

With the advent of massive social and information networks, link-based clustering of objects in networks becomes increasingly important since it may help discover hidden knowledge in large networks. Link-based clustering groups objects based on their links instead of attribute values. This is especially useful when attributes of objects cannot be fully obtained. Most existing link-based clustering algorithms are on homogeneous networks, where links carry the same semantic meaning and only differ in their strengths (i.e., weights). However, most real-world networks are heterogeneous, where objects are of multiple types and are linked via different types of relations or sequences of relations, forming a set of *meta-paths*. These meta-paths indicate different relations among object types and imply diverse semantics, and thus clustering on different meta-paths will generate rather different results, as shown below.

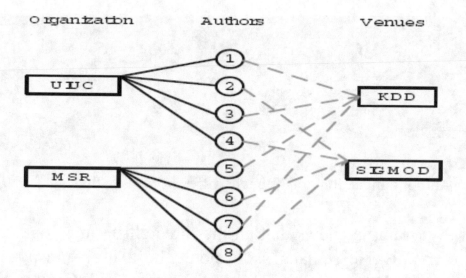

Figure 7.1: A toy heterogeneous information network containing organizations, authors, and venues.

Example 7.1 (Meta-path-based clustering) A toy heterogeneous information network is shown in Figure 7.1, which contains three types of objects: organization (O), author (A), and venue (V), and two types of links: the solid line represents the affiliation relation between author and organization, whereas the dashed one the publication relation between author and venue. Authors are then connected (indirectly) via different meta-paths. For example, $A - O - A$ is a meta-path denoting a relation between authors via organizations (i.e., colleagues), whereas $A - V - A$ denotes a relation between authors via venues (i.e., publishing in the same venues). A question then arises: *which type of connections should we use to cluster the authors?*

Obviously, there is no unique answer to this question: Different meta-paths lead to different author connection graphs, which may lead to different clustering results. In Figure 7.2(a), authors are connected via organizations and form two clusters: {1, 2, 3, 4} and {5, 6, 7, 8}; in Figure 7.2(b), authors are connected via venues and form two different clusters: {1, 3, 5, 7} and {2, 4, 6, 8}; whereas in Figure 7.2(c), a connection graph combining both meta-paths generates 4 clusters: {1, 3}, {2, 4}, {5, 7}, and {6, 8}.

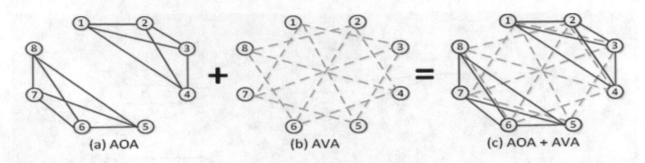

Figure 7.2: Author connection graphs under different meta-paths.

This toy example shows that all the three clusterings look reasonable but they carry diverse semantics. It should be a user's responsibility to choose her desired meta-path(s). However, it is often difficult to ask her to explicitly specify one or a weighted combination of meta-paths. Instead, it is easier for her to give some guidance in other forms, such as giving one or a couple of examples for each cluster. For example, it may not be hard to give a few known conferences in each cluster (i.e., field) if one wants to cluster them into K research areas (for a user-desired K), or ask a user to name a few restaurants if one wants to cluster them into different categories in a business review website (e.g., Yelp).

The new situation is that since we are dealing with heterogeneous networks, the previous work on user-guided clustering or semi-supervised learning approaches on (homogeneous) graphs [36; 88; 89] cannot apply. We need to explore meta-paths that represent heterogeneous connections across objects, leading to rich semantic meanings, hence diverse clustering results. With user guidance, a system will be able to learn the most appropriate meta-paths or their weighted combinations. The learned meta-paths will in turn provide an insightful view to help understand the underlying mechanism in the formation of a specific type of clustering, such as which meta-path is more important to determine a restaurant's category?—the meta-path connecting them via customers, the one connecting them via text in reviews, or the one determined by the nearest spatial locations?

We thus integrate meta-path selection with user-guided clustering in order to better cluster a user-specified type of objects (i.e., *target objects*) in a heterogeneous information network. We assume that user guidance is in the form of a small set of seeds for each cluster. For example, to cluster authors into two clusters in Example 7.1, a user may seed $\{1\}$ and $\{5\}$ for two clusters, which implies a selection of meta-path $A - O - A$; or seed $\{1\}$, $\{2\}$, $\{5\}$, and $\{6\}$ for four clusters, which implies a combination of both meta-paths $A - O - A$ and $A - V - A$ with about equal weight. Our goal is to (1) determine the weight of each meta-path for a particular clustering task, which should be consistent with the clustering results implied by the limited user guidance, and (2) output the clustering result according to the user guidance and using the learned weights for each meta-path.

We propose a probabilistic model that models the hidden clusters for target objects, the user guidance, and the quality weights for different meta-paths in a unified framework. An effective and efficient iterative algorithm *PathSelClus* is developed to learn the model, where the clustering quality and the meta-paths quality mutually enhance each other. The experiments with different tasks on two real networks show our algorithm outperforms the baselines.

7.2 THE META-PATH SELECTION PROBLEM FOR USER-GUIDED CLUSTERING

Here we illustrate the problem using two heterogeneous information networks: the DBLP network and the Yelp network.

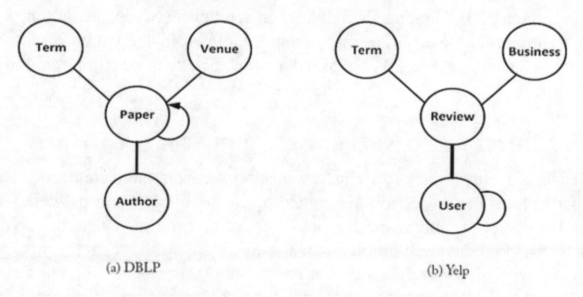

(a) DBLP   (b) Yelp

Figure 7.3: Examples of heterogeneous information networks.

Example 7.2 (The DBLP bibliographic network) As introduced before, DBLP is a computer science bibliographic network (see schema in Figure 7.3(a)) containing 4 types of objects: **paper**(P), **author** (A), **term** (T), and **venue** (V) (i.e., conferences and journals). Links exist between authors and papers by the relation of "write" and "written by," between papers and terms by "mention" and "mentioned by," and between venues and papers by "publish" and "published by." "Citation" relation between papers can be added further using other data source, such as Google scholar.

Example 7.3 (The Yelp network) Yelp is a website (http://www.yelp.com/) where users can write reviews for businesses. The Yelp network (see schema in Figure 7.3(b)) used in this chapter contains 4 types of objects: **business** (B), **user** (U), **term** (T), and **review** (R). Links exist between users and reviews by the relation of "write" and "written by," between reviews and terms by "mention" and "mentioned by," between businesses and reviews by "commented by" and "comment," and between users by "friendship" (not included in our dataset).

Following our previous discussion, a meta-path is defined by a sequence of relations in the network schema and can be denoted by a sequence of object types when there is no ambiguity. For example, $A - P - A$ is a meta-path denoting the co-authorship between authors, and $A - P - V$ is a meta-path denoting the publication relation between the author and the venue type. Note that a single relation defined in the network schema can be viewed as a special case of meta-path, such as the citation relation $P \rightarrow P$.

7.2.1 THE META-PATH SELECTION PROBLEM

Link-based clustering is to cluster objects based on their connections to other objects in the network. In a heterogeneous information network, we need to specify more information for a meaningful clustering. This includes: (1) the type of objects to be clustered (called the **target type**); and (2) the type of connections, that is, meta-path(s), to use for the clustering task, and we call the object type that the target type is connecting to via the meta-path as the **feature type**. For example, when clustering authors based on the venues which they have published papers in, the target type is the author type, the meta-path to use is $A - P - V$, and the feature type is venue.

In a heterogeneous information network, target objects could link to many types of feature objects by multiple meta-paths. For example, authors could connect to other authors via meta-path $A - P - A$, or connect to terms via meta-path $A - P - T$. Meta-path selection is to determine which meta-paths or their weighted combination to use for a specific clustering task.

7.2.2 USER-GUIDED CLUSTERING

User guidance is critical for clustering objects in the network. In this study, we consider the guidance as user seeding objects in each cluster. For example, to cluster authors based on their (hidden) research areas, one can first provide several representative authors in each area. These seeds are used as guidance for clustering all the target objects in the network. More importantly, they provide information for selecting the most relevant meta-paths for the specific clustering task. Note that in practice, a user may not be able to provide seeds for *every* cluster, but only for *some* clusters they are most familiar with, which should be handled by the algorithm too.

7.2.3 THE PROBLEM DEFINITION

Now we provide the problem definition of user-guided clustering via meta-path selection. Given a heterogeneous information network G, a user needs to specify the following as inputs for a clustering task.

1. The target type for clustering, type T.

2. The number of clusters, K, and the object seeds for each cluster, say $\mathcal{L}_1, \ldots, \mathcal{L}_K$, where $\mathcal{L}_k$ denotes the object seeds for cluster k, which could be an empty set. These seeds will be used as hints to learn the purpose/preference of the clustering task.

3. A set of M meta-paths starting from type T, denoted as $\mathcal{P}_1, \mathcal{P}_2, \ldots, \mathcal{P}_M$, which might be helpful for the clustering task. These meta-paths can be determined either according to users' expert knowledge, or by traversing the network schema starting from type T with a length constraint.

For each meta-path $\mathcal{P}_m$, we calculate the adjacency matrix W_m, which we call *relation matrix*, between the target type T and the feature type F_m, by multiplying adjacency matrices for each relation along the meta-path. For example, the relation matrix W for meta-path $A - P - V$, denoting the number of papers published by an author in a venue, is calculated by $W = W_{AP} \times W_{PV}$, where W_{AP} and W_{PV} are the adjacency matrices for relation $A - P$ and $P - V$, respectively.

The output of the algorithm consists of two parts: (1) the weight $\alpha_m \geq 0$ of each meta-path $\mathcal{P}_m$ for a particular clustering task, which should be consistent with the clustering result implied by the limited user guidance; and (2) the clustering result according to the user guidance and under the learned weights for each meta-path, that is, to associate each target object t_i in T with a K-dimensional soft clustering probability vector, $\boldsymbol{\theta}_i = (\theta_{i1}, \ldots, \theta_{iK})$, where θ_{ik} is the probability of t_i belonging to cluster k, i.e., $\theta_{ik} \geq 0$ and $\sum_{k=1}^{K} \theta_{ik} = 1$.

7.3 THE PROBABILISTIC MODEL

A good clustering result is determined by several factors: First, the clustering result should be consistent with the link structure; second, it should also be consistent with the user guidance; and third, the importance of each meta-path is implied by the user-guided clustering, which should be modeled and learned to further enhance the clustering quality. In the following, we propose a probabilistic approach to model the problem in a unified framework, by considering all the three factors.

7.3.1 MODELING THE RELATIONSHIP GENERATION

To model the consistency between a clustering result and a relation matrix, we propose a clustering-based generative model for relationship generation.

For a meta-path $\mathcal{P}_m$, let its corresponding relation matrix between the target type T and the feature type F_m be W_m. For each target object t_i, we model its relationships as generated from a mixture of multinomial distributions, where the probability of $t_i \in T$ connecting to $f_{j,m} \in F_m$ is conditionally independent on t_i given that the hidden cluster label of the relationship is known. Let $\pi_{ij,m} = P(j|i, m)$ be the generative probability of the relationship starting from t_i and ending at $f_{j,m}$, where $\sum_j \pi_{ij,m} = 1$, then

$$\pi_{ij,m} = P(j|i, m) = \sum_k P(k|i) P(j|k, m) = \sum_k \theta_{ik} \beta_{kj,m} \,, \tag{7.1}$$

where $\theta_{ik} = P(k|i)$ denotes the probability of t_i belonging to cluster k and $\beta_{kj,m} = P(j|k, m)$ denotes the probability of $f_{j,m}$ appearing in cluster k. In other words, let $\boldsymbol{\pi}_{i,m} = (\pi_{i1,m}, \ldots, \pi_{i|Fm|,m})$ be the generative probability vector for target object t_i, then each $\boldsymbol{\pi}_{i,m}$ can be factorized as a weighted summation of ranking distributions of feature objects in each cluster. The factorization idea is similar to that of PLSA [26], PHITS [15], and RankClus [66], but is built on meta-path-encoded relationships rather than immediate links. This extension will capture more and richer link-based features for clustering target objects in heterogeneous networks.

By assuming each target object t_i is independent with each other and each relationship generated by t_i is independent with each other, the probability of observing all the relationships between all the target objects and feature objects is the production of the probability of all the relationships following meta-path $\mathcal{P}_m$:

$$P(W_m|\Pi_m, \Theta, B_m) = \prod_i P(\mathbf{w}_{i,m}|\boldsymbol{\pi}_{i,m}, \Theta, B_m) = \prod_i \prod_j (\pi_{ij,m})^{w_{ij,m}} \,, \tag{7.2}$$

where $\Pi_m = \Theta B_m$ is the probability matrix with cells as $\pi_{ij,m}$'s, Θ is the parameter matrix for θ_{ik}'s, B_m is the parameter matrix for $\beta_{kj,m}$'s, and $w_{ij,m}$ is the weight of the relationship between t_i and $f_{j,m}$. Note that, to model the relationship generation, each meta-path $\mathcal{P}_m$ corresponds to a different generative probability matrix Π_m. These probability matrices share the same soft clustering probabilities Θ, but they have different ranking distributions B_m in different meta-paths.

7.3.2 MODELING THE GUIDANCE FROM USERS

Further, we take the user guidance in the form of object seeds for some clusters as the prior knowledge for the clustering result Θ, by modeling the prior as a Dirichlet distribution rather than treating them as hard labeled ones.

For each target object t_i, its clustering probability vector θ_i is assumed to be a multinomial distribution, which is generated from some Dirichlet distribution. If t_i is labeled as a seed in cluster k^*, θ_i is then modeled as being sampled from a Dirichlet distribution with parameter vector $\lambda_{ek^*} + \mathbf{1}$, where $\mathbf{e}_{k^*}$ is a K-dimensional basis vector, with the k^*th element as 1 and 0 elsewhere. If t_i is not a seed, θ_i is then assumed as being sampled from a uniform distribution, which can also be viewed as a Dirichlet distribution with parameter vector $\mathbf{1}$. The density of θ_i given such priors is:

$$P(\theta_i|\lambda) \propto \begin{cases} \prod_k \theta_{ik}^{\mathbf{1}_{\{t_i \in \mathcal{L}_k\}}\lambda} = \theta_{ik^*}^{\lambda}, & \text{if } t_i \text{ is labeled and } t_i \in \mathcal{L}_{k^*}, \\ 1, & \text{if } t_i \text{ is not labeled,} \end{cases} \tag{7.3}$$

where $\mathbf{1}_{\{ti \in \mathcal{L}_k\}}$ is an indicator function, which is 1 if $t_i \in \mathcal{L}_k$ holds, and 0 otherwise.

The hyper-parameter λ is a non-negative value, which controls the strength of users' confidence over the object seeds in each cluster. From Equation (7.3), we can find that:

- when $\lambda = 0$, the prior for θ_i of a labeled target object becomes a uniform distribution, which means no guidance information will be used in the clustering process;

- when $\lambda \to \infty$, the prior for θ_i of a labeled target object converges to a point mass, i.e., $P(\theta_i = \mathbf{e}_{k^*}) \to 1$ or $\theta_i \to \mathbf{e}_{k^*}$, which means we will assign k^* as the hard cluster label for t_i.

In general, a larger λ indicates a higher probability that θ_i is around the point mass $\mathbf{e}_{k^*}$, and thus a higher confidence for the user guidance.

7.3.3 MODELING THE QUALITY WEIGHTS FOR META-PATH SELECTION

Different meta-paths may lead to different clustering results, therefore it is desirable to learn the quality of each meta-path for the specific clustering task. We propose to learn the quality weight for each meta-path by evaluating the consistency between its relation matrix and the user-guided clustering result.

In deciding the clustering result for target objects, a meta-path may be of low quality for the following reasons.

1. The relation matrix derived by the meta-path does not contain an inherent cluster structure. For example, target objects are connecting to the feature objects randomly.

2. The relation matrix derived by the meta-path itself has a good inherent cluster structure, however, it is not consistent with the user guidance. For example, in our motivating example, if the user gives a guidance as: $K = 2$, $\mathcal{L}_1 = \{1\}$, $\mathcal{L}_2 = \{2\}$, then the meta-path $A - O - A$ should have a lower impact in the clustering process for authors.

The general idea of measuring the quality of each meta-path is to see whether the relation matrix W_m is consistent with the detected hidden clusters Θ and thus the generative probability matrix Π_m, which is a function of Θ, i.e., $\Pi_m = \Theta\, B_m$.

In order to quantify the weight for such quality, we model the weight α_m for meta-path $\mathcal{P}_m$ as the *relative weight* for each relationship between target objects and feature objects following $\mathcal{P}_m$. In other words, we treat our observations of the relation matrix as $\alpha_m\, W_m$ rather than original W_m. A larger α_m indicates a higher quality and a higher confidence of the observed relationships, and thus each relationship should count more.

Then, we assume the multinomial distribution $\pi_{i,m}$ has a prior of Dirichlet distribution with parameter vector ϕ_i. In particular, we consider a discrete uniform prior, which is a special case of Dirichlet distribution with parameters as an all-one vector, i.e., $\phi_{i,m} = \mathbf{1}$. The value of α_m is determined by the consistency between the observed relation matrix W_m and the generative probability matrix Π_m, which can be evaluated as how likely we can get Π_m given the relation matrix W_m and its quality weight α_m. The goal is then to find the α_m^* that maximizes the posterior probability of $\pi_{i,m}$ for all the target objects t_i, given the observation of relationships $\mathbf{w}_{i,m}$ with relative weight α_m:

$$\alpha_m^* = \arg\max_{\alpha_m} \prod_i P(\pi_{i,m}|\alpha_m \mathbf{w}_{i,m}, \theta_i, B_m) . \tag{7.4}$$

We can show that the posterior of $\pi_{i,m} = \theta_i B_m$ is another Dirichlet distribution with the updated parameter vector as $\alpha_m \mathbf{w}_{i,m} + \mathbf{1}$, according to the multinomial-Dirichlet conjugate:

$$\pi_{i,m}|\alpha_m \mathbf{w}_{i,m}, \theta_i, B_m \sim Dir(\alpha_m w_{ij,m} + 1, \ldots, \alpha_m w_{i|F_m|,m} + 1) \tag{7.5}$$

which has the following density function:

$$P(\pi_{i,m}|\alpha_m \mathbf{w}_{i,m}, \theta_i, B_m) = \frac{\Gamma(\alpha_m n_{i,m} + |F_m|)}{\prod_j \Gamma(\alpha_m w_{ij,m} + 1)} \prod_j (\pi_{ij,m})^{\alpha_m w_{ij,m}} , \tag{7.6}$$

where $n_{i,m} = \sum_j w_{i,j,m}$, the total number of path instances from t_i following meta-path $\mathcal{P}_m$. By modeling α_m in such a way, the meaning of α_m is quite clear.

- $\alpha_m w_{ij,m} + 1$ is the parameter of jth dimension for the new Dirichlet distribution.

- The larger α_m, the more likely it will generate a $\pi_{i,m}$ with a distribution as the observed relationship distribution, i.e., $\pi_{i,m} \to \mathbf{w}_{i,m}/n_{i,m}$ when $\alpha_m \to \infty$, where $n_{i,m}$ is the total number of path instances from t_i following meta-path $\mathcal{P}_m$.

- The smaller α_m, the more likely it will generate a π_i with a uniform distribution (which means randomly), i.e., $\pi_{i,m} \to (1/|F_m|, \ldots, 1/|F_m|)$ when $\alpha_m \to 0$, where $|F_m|$ is the total size of feature objects in meta-path $\mathcal{P}_m$.

Note that we do not consider negative α_m's in this model, which means that the relationships with a negative impact in the clustering process are not considered, and the extreme case of $\alpha_m = 0$ means that the relationships in a meta-path are totally irrelevant to the clustering process.

7.3.4 THE UNIFIED MODEL

Putting all the three factors together, we have the joint probability of observing the relation matrices with relative weights α_m's, and the parameter matrices Π_m's and Θ:

$$P(\{\alpha_m W_m\}_{m=1}^M, \Pi_{1:M}, \Theta | B_{1:M}, \Phi_{1:M}, \lambda)$$
$$= \prod_i (\prod_m P(\alpha_m W_m | \Pi_m, \theta_i, B_m) P(\Pi_m | \Phi_m)) P(\theta_i | \lambda), \qquad (7.7)$$

where Φ_m is the Dirichlet prior parameter matrix for Π_m, and an all-one matrix in our case. We want to find the maximum a posteriori probability (MAP) estimate for Π_m's and Θ, which maximizes the logarithm of posterior probability of $\{\Pi_m\}_{m=1}^M$, given the observations of relation matrices with relative weights $\{\alpha_m W_m\}_{m=1}^M$ and Θ, plus a regularization term over θ_i for each target object denoting the logarithm of prior density of θ_i:

$$J = \sum_i (\sum_m \log P(\pi_{i,m} | \alpha_m w_{i,m}, \theta_i, B_m) + \sum_k 1_{\{l_i \in \mathcal{L}_k\}} \lambda \log \theta_{ik}). \qquad (7.8)$$

By substituting the posterior probability formula in Equation (7.6) and the factorization form for all $\pi_{i,m}$, we get the final objective function:

$$J = \sum_i \left(\sum_m \left(\sum_j \alpha_m w_{ij,m} \log \sum_k \theta_{ik} \beta_{kj,m} \right. \right.$$

$$+ \log \Gamma(\alpha_m n_{i,m} + |F_m|) - \sum_j \log \Gamma(\alpha_m w_{ij,m} + 1)) \tag{7.9}$$

$$\left. + \sum_k \mathbf{1}_{\{t_i \in \mathcal{L}_k\}} \lambda \log \theta_{ik} \right) .$$

7.4 THE LEARNING ALGORITHM

In this section, we introduce the learning algorithm, *PathSelClus*, for the model (Equation (7.9)) proposed in Section 7.3. It is a two-step iterative algorithm, where the clustering result Θ and the weights for each meta-path α mutually enhance each other. In the first step, we fix the weight vector α, and learn the best clustering results Θ under this weight. In the second step, we fix the clustering matrix Θ and learn the best weight vector α.

7.4.1 OPTIMIZE CLUSTERING RESULT GIVEN META-PATH WEIGHTS

When α is fixed, the terms only involving α can be discarded in the objective function Equation (7.9), which is then reduced to:

$$J_1 = \sum_m \alpha_m \sum_i \sum_j w_{ij,m} \log \sum_k \theta_{ik} \beta_{kj,m} + \sum_i \sum_k \mathbf{1}_{\{t_i \in \mathcal{L}_k\}} \lambda \log \theta_{ik} . \tag{7.10}$$

The new objective function can be viewed as a weighted summation of the log-likelihood for each relation matrix under each meta-path, where the weight α_m indicates the quality of each meta-path, plus a regularization term over Θ representing the user guidance. Θ and the augmented parameter B_m's can be learned using the standard EM algorithm, as follows.

• **E-step:** In each relation matrix, we use $z_{ij,m}$ to denote the cluster label for each relationship between a target object t_i and a feature object $f_{j,m}$. According to the generative process described in Section 7.3.1, a cluster k is first picked with probability θ_{ik}, and a feature object $f_{j,m}$ is picked with probability $\beta_{kj,m}$. The conditional probability of the hidden cluster label given the old Θ^{t-1} and B_m^{t-1} values is then:

$$p(z_{ij,m} = k | \Theta^{t-1}, B_m^{t-1}) \propto \theta_{ik}^{t-1} \beta_{kj,m}^{t-1}. \tag{7.11}$$

- **M-step:** We have the updating formulas for Θ^t and B_m^t as:

$$\theta_{ik}^t \propto \sum_m \alpha_m \sum_j w_{ij,m} p(z_{ij,m} = k | \Theta^{t-1}, B_m^{t-1}) + 1_{\{t_i \in \mathcal{L}_k\}} \lambda \tag{7.12}$$

$$\beta_{kj,m}^t \propto \sum_i \sum_j w_{ij,m} p(z_{ij,m} = k | \Theta^{t-1}, B_m^{t-1}). \tag{7.13}$$

From Equation (7.12), we can see that the clustering membership vector θ_i for t_i is determined by the cluster labels of its relationships to all the feature objects in all the relation matrices. Besides, if t_i is labeled as a seed object in some cluster $k*$, θ_i is also determined by the label. The strength of impacts from these factors is determined by the weight of each meta-path α_m, and the strength of the cluster labels λ, where α_m's are learned automatically by our algorithm, and λ is given by users.

7.4.2 OPTIMIZE META-PATH WEIGHTS GIVEN CLUSTERING RESULT

Once a clustering result Θ and the augmented parameter B_m's are given, we can calculate the generative probability matrix Π_m for each meta-path $\mathcal{P}_m$ by: $\Pi_m = \Theta B_m$. By discarding the irrelevant terms, the objective function of Equation (7.9) can be reduced to:

$$J_2 = \sum_i \left(\sum_m \left(\sum_j \alpha_m w_{ij,m} \log \pi_{ij,m} + \log \Gamma(\alpha_m n_{i,m} + |F_m|) - \sum_j \log \Gamma(\alpha_m w_{ij,m} + 1) \right) \right). \tag{7.14}$$

It is easy to check that J_2 is a concave function, which means there is a unique α that maximizes J_2. We use gradient descent approach to solve the problem, which is an iterative algorithm with the updating formula as:

$\alpha_m^t = \alpha_m^{t-1} + \eta_m^t \left. \frac{\partial J_2}{\partial \alpha_m} \right|_{\alpha_m = \alpha_m^{t-1}}$. To guarantee the increase of J_2, the step size η_m^t is usually set as a small enough number. By setting $\eta_m^t = \frac{\alpha_m^{t-1}}{-\sum_i \sum_j w_{ij,m} \log \pi_{ij,m}}$, following the trick used in non-negative matrix factorization (NMF) [37], we can get updating formula for α_m as:

$$\alpha_m^t = \alpha_m^{t-1} \frac{\sum_i \left(\psi(\alpha_m^{t-1} n_{im} + |F_m|) n_{i,m} - \sum_j \psi(\alpha_m^{t-1} w_{ij,m} + 1) w_{ij,m} \right)}{-\sum_i \sum_j w_{ij,m} \log \pi_{ij,m}} \qquad (7.15)$$

which guarantees α_m^t a *non-negative* value, where $\psi(x)$ is the digamma function, the first derivative of $\log \Gamma (x)$. Also, by looking at the denominator of the formula, we can see that a larger log-likelihood of observing relationships $w_{ij,m}$ under model probability $\pi_{ij,m}$ (i.e., a smaller denominator as log-likelihood is negative) generally leads to a larger α_m. This is also consistent with the human intuition.

7.4.3 THE PATHSELCLUS ALGORITHM

Overall, the *PathSelClus* algorithm is an iterative algorithm that optimizes Θ and α alternatively. The optimization of Θ contains an inner loop of EM-algorithm, and the optimization of α contains another inner loop of gradient descent algorithm.

The Weight Setting of Relation Matrices Given a heterogeneous information network G, we calculate the relation matrix W_m for each given meta-path $\mathcal{P}_m$ by multiplying adjacency matrices along the meta-path. It can be shown that, scaling W_m by a factor of $1/c_m$ leads to a scaling of the learned relative weight α_m by a factor of c_m. Therefore, the performance of the clustering result will not be affected by the scaling of the relation matrix, which is a good property of our algorithm.

Initialization Issues. For the initial value of α, we set it as an all-one vector, which assumes all the meta-paths are equally important. For the initial value of Θ in the clustering step given α, if t_i is not labeled, we assign a random clustering vector to $\boldsymbol{\theta}_i$, whereas if t_i is labeled as a seed for a cluster k^*, we assign $\boldsymbol{\theta}_i = \mathbf{e}_k^*$.

Time Complexity Analysis. The *PathSelClus* algorithm is very efficient, as it is proportional to the number of relationships that are used in the clustering process, which is about linear to the number of target objects for short meta-paths in *sparse* networks. Formally, for the inner EM algorithm that optimizes Θ, the time complexity is $O(t_1(K \sum_m |\mathcal{E}_m| + K|T| + K \sum_m |F_m|))$ $= O(t_1(K \sum_m |\mathcal{E}_m|))$, where $|\mathcal{E}_m|$ is the number of non-empty relationships in relation matrix W_m, $|T|$ and $|F_m|$ are the numbers of target objects and feature objects in meta-path $\mathcal{P}_m$, which are typically smaller than $|\mathcal{E}_m|$, and t_1 is the number of iterations. For the inner gradient descent algorithm, the time complexity is $O(t_2(\sum_m |\mathcal{E}_m|))$, where t_2 is the number of iterations. The total time complexity for the whole algorithm is then $O(t(t_1(K \sum_m |\mathcal{E}_m|) + t_2(\sum_m |\mathcal{E}_m|)))$, where t is the number of outer iterations, which usually is a small number. Such a processing efficiency has also been verified by our experiments.

7.5 EXPERIMENTAL RESULTS

In this section, we compare *PathSelClus* with several baselines and show the effectiveness of our algorithm.

7.5.1 DATASETS

We use two real information networks for performance test, the DBLP network and the Yelp network. For each network, we design multiple clustering tasks provided with different user guidance, which are introduced in the following.

1. **The DBLP Network.** For the DBLP network introduced early in the chapter, we design three clustering tasks in the following.

    • DBLP-T1: Cluster conferences in the *"four-area"* dataset [69], which contains 20 major conferences and all the related papers, authors and terms in DM, DB, IR, and AI fields, according to the *research areas* of the conferences. The candidate meta-paths include: $V - P - A - P - V$ and $V - P - T - P - V$.

- DBLP-T2: Cluster top-2000 authors (by their number of publications) in the *"four-area" dataset*, according to their *research areas*. The candidate meta-paths include: $A - P - A$, $A - P - A - P - A$, $A - P - V - P - A$, and $A - P - T - P - A$.

- DBLP-T3: Cluster 165 authors who have been ever advised by Christos Faloutsos, Michael I. Jordan, Jiawei Han, and Dan Roth (including these professors), according to their *research groups*. The candidate meta-paths are the same as in DBLP-T2.

2. **The Yelp Network.** For the Yelp network introduced early in the chapter, we are provided by Yelp a sub-network[1], which include 6900 businesses, 152,327 reviews, and 65,888 users. Hierarchical categories are provided for each business as well, such as "Restaurants," "Shopping," and so on. For Yelp network, we design three clustering tasks in the following.

- Yelp-T1: We select 4 relatively big categories ("Health and Medical," "Food," "Shopping," and "Beauty and Spas"), and cluster 2224 businesses with more than one reviews according to two meta-paths: $B - R - U - R - B$ and $B - R - T - R - B$.

- Yelp-T2: We select 6 relatively big sub-categories under the first-level category "Restaurant" ("Sandwiches," "Thai," "American (New)," "Mexican," "Italian," and "Chinese"), and cluster 554 businesses with more than one reviews according to the same two meta-paths.

- Yelp-T3: We select 6 relatively big sub-categories under the first-level category "Shopping" ("Eyewear & Opticians," "Books, Mags, Music and Video," "Sporting Goods," "Fashion," "Drugstores," and "Home & Garden"), and cluster 484 businesses with more than one reviews according to the same two meta-paths.

7.5.2 EFFECTIVENESS STUDY

First, we study the effectiveness of our algorithm under different tasks, and compare it with several baselines.

Baselines

Three baselines are used for comparison studies. Since none of them has considered the meta-path selection problem, we will use all the meta-paths as features and prepare them to fit the input of each of these algorithms. The first is user-guided, information theoretic-based, k-means clustering (ITC), which is an adaption of seeded k-means algorithm proposed in [6], by replacing Euclidean distance to KL-divergence as used in information theoretic-based clustering algorithms [3; 18]. ITC is a hard clustering algorithm. For the input, we concatenate all the relation matrices side-by-side into one single relation matrix, and thus we get a very high dimensional feature vector for each target object.

The second baseline is the label propagation (LP) algorithm proposed in [89], which utilizes link structure to propagate labels to the rest of the network. For the input, we add all the relation matrices together to get one single relation matrix. As LP is designed for homogeneous networks, we confine our meta-paths to ones that start and end both in the target type. LP is a soft clustering algorithm.

The third baseline is the cluster ensemble algorithm proposed in [53], which can combine soft clustering results into a consensus, which we call ensemble_soft. Different from the previous two baselines that directly combine meta-paths at the input level, cluster ensemble combines the clustering results for different meta-paths at the output level. Besides, we also use majority voting as another baseline (ensemble_voting), which first maps each clustering result for each target object into a hard cluster label and then picks the cluster label that is the majority over different meta-paths. As we can use either ITC or LP as the clustering algorithm for each ensemble method, we get four ensemble baselines in total: ITC_soft, ITC_voting, LP_soft, and LP_voting.

Evaluation Methods

Two evaluation methods are used to test the clustering result compared with the ground truth, where the soft clustering is mapped into hard cluster labels. The first measure is *accuracy*, which is used when seeds are available for every cluster and is calculated as the percentage of target objects going to the correct cluster. Note that, in order to measure whether the seeds are indeed attracting objects to the right cluster, we do not map the outcome cluster labels to the given class labels. The second measure is *normalized mutual information (NMI)*, which does not require the mapping relation between ground truth labels and the cluster labels obtained by the clustering algorithm. Both measures are in the range of 0 to 1, and a higher value indicates a better clustering result in terms of the ground truth.

Full Cluster Seeds

We first test the clustering accuracy when cluster seeds are given for every cluster. In this case, all the three baselines can be used and compared. Performances under different numbers of seeds in each cluster are tested. Each result is the average of 10 runs.

The accuracy for all the 6 tasks for two networks are summarized in Tables 7.1–7.3 and Tables 7.4–7.6, respectively. From the results we can see that, *PathSelClus* performs the best in most of the tasks. Even for the task such as DBLP-T3 where other methods give the best clustering result, *PathSelClus* still gives clustering results among the top. This means, *PathSelClus* can give consistently good results across different tasks in different networks.

Table 7.1: Clustering accuracy for DBLP-T1 task

#S	Measure	PathSelClus	LP	ITC	LP_voting	LP_soft	ITC_voting	ITC_soft
1	Accuracy	**0.9950**	0.6500	0.6900	0.6500	0.6650	0.6450	0.5100
1	NMI	**0.9906**	0.6181	0.6986	0.6181	0.5801	0.5903	0.5316
2	Accuracy	1	0.7500	0.8450	0.7500	0.8200	0.8950	0.8700
2	NMI	1	0.6734	0.7752	0.6734	0.7492	0.8321	0.7942

Table 7.2: Clustering accuracy for DBLP-T2 task

#S	Measure	PathSelClus	LP	ITC	LP_voting	LP_soft	ITC_voting	ITC_soft
1	Accuracy	**0.7951**	0.2122	0.3284	0.2109	0.3529	0.2513	0.2548
	NMI	**0.6770**	0.0312	0.1277	0.0267	0.0301	0.4317	0.4398
5	Accuracy	**0.8815**	0.2487	0.3223	0.5117	0.3685	0.3311	0.3495
	NMI	**0.6868**	0.0991	0.1102	0.4402	0.0760	0.3092	0.4316
10	Accuracy	**0.8863**	0.5586	0.3694	0.4297	0.3880	0.4891	0.2969
	NMI	**0.6947**	0.4025	0.1261	0.1788	0.1148	0.4045	0.4204

Table 7.3: Clustering accuracy for DBLP-T3 task

#S	Measure	PathSelClus	LP	ITC	LP_voting	LP_soft	ITC_voting	ITC_soft
1	Accuracy	0.8067	**0.9273**	0.5376	0.7091	0.5424	0.4770	0.2358
	NMI	0.6050	**0.7966**	0.5120	0.5870	0.7182	0.3008	0.3416
2	Accuracy	0.9036	**0.9394**	0.5285	0.7333	0.3267	0.5176	0.4085
	NMI	0.7485	**0.8283**	0.5056	0.5986	0.8087	0.3898	0.3464
4	Accuracy	0.9248	**0.9576**	0.7624	0.7636	0.9255	0.6370	0.5485
	NMI	0.7933	0.8841	0.6280	0.6179	**0.9057**	0.4437	0.4634

Also, by looking at the clustering accuracy trend along with the number of seeds used in each cluster, we can see that more seeds generally leads to better clustering results.

Table 7.4: Clustering accuracy for Yelp-T1 task

%S	Measure	PathSelClus	LP	ITC	LP_voting	LP_soft	ITC_voting	ITC_soft
1%	Accuracy	**0.5384**	0.3381	0.2619	0.1632	0.1632	0.2564	0.2769
	NMI	**0.5826**	0.0393	0.0042	0.0399	0.0399	0.1907	0.2435
2%	Accuracy	**0.5487**	0.3444	0.2798	0.1713	0.1713	0.3581	0.3790
	NMI	**0.5800**	0.0557	0.0062	0.0567	0.0567	0.2281	0.2734
5%	Accuracy	**0.5989**	0.3732	0.3136	0.1965	0.1965	0.5215	0.5250
	NMI	**0.5796**	0.1004	0.0098	0.0962	0.0962	0.2583	0.2878

Table 7.5: Clustering accuracy for Yelp-T2 task

%S	Measure	PathSelClus	LP	ITC	LP_voting	LP_soft	ITC_voting	ITC_soft
1%	Accuracy	**0.7435**	0.1137	0.1758	0.2112	0.2112	0.2430	0.2022
	NMI	**0.6517**	0.0323	0.0178	0.0578	0.0578	0.2308	0.2490
2%	Accuracy	**0.8004**	0.1264	0.1910	0.2202	0.2202	0.2762	0.2792
	NMI	**0.6803**	0.0487	0.0150	0.0801	0.0801	0.2099	0.2907
5%	Accuracy	**0.8125**	0.2653	0.2200	0.2437	0.2437	0.3049	0.3240
	NMI	**0.6894**	0.1111	0.0220	0.1212	0.1212	0.2252	0.2692

Table 7.6: Clustering accuracy for Yelp-T3 task

%S	Measure	PathSelClus	LP	ITC	LP_voting	LP_soft	ITC_voting	ITC_soft
1%	Accuracy	**0.4736**	0.2789	0.1893	0.0682	0.0682	0.2593	0.1775
	NMI	**0.4304**	0.0568	0.0155	0.0626	0.0626	0.1738	0.2065
2%	Accuracy	**0.4597**	0.4008	0.1948	0.0764	0.0764	0.2318	0.2033
	NMI	**0.4359**	0.0910	0.0172	0.0755	0.0755	0.1835	0.1822
5%	Accuracy	0.4393	**0.5351**	0.2233	0.1033	0.1033	0.3337	0.3083
	NMI	**0.4415**	0.1761	0.0194	0.1133	0.1133	0.1793	0.2285

Partial Cluster Seeds

We then test the clustering accuracy when cluster seeds are only available for some of the clusters. We perform this study on DBLP-T3 using *PathSelClus*, which includes four clusters, and the results are shown in Figure 7.4. We can see that even if user guidance is only given to some clusters, those seeds can still be used to improve the clustering accuracy. In general, the fewer number of clusters with seeds, the worse the clustering accuracy, which is consistent with the human intuition. Note that label propagation-based methods like LP cannot deal with partial cluster labels. However, in reality it is quite common that users are only familiar with some of the clusters and are only able to give good seeds in those clusters. That is another advantage of *PathSelClus*.

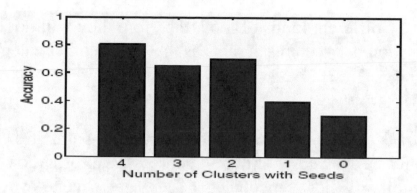

Figure 7.4: Clustering accuracy under partial guidance on DBLP-T3. Number of seeds provided by user for each cluster is 1 (#seeds = 1).

7.5.3 CASE STUDY ON META-PATH WEIGHTS

One of the major contributions of *PathSelClus* is that it can select the right meta-paths for a user-guided clustering task. We now show the learned weights of meta-paths for some of the tasks.

In DBLP-T1 task, the total weight α_m for meta-path $V - P - A - P - V$ is 1576, and the average weight per relationship (a concrete path instance following the meta-path) is 0.0017. The total weight for meta-path $V - P - T - P - V$ is 17,001, whereas the average weight per relationship is 0.0003. This means that generally the relationships between two conferences that are connected by an author are more trustable than the ones that are connected by a term, which is consistent with human intuition since many terms can be used in different research areas and authors are typically more focused on confined research topics. However, as there are much more relationships following $V - P - T - P - V$ than following $V - P - A - P - V$, the former overall provide more information for clustering.

In the Yelp network, similar to DBLP-T1 task, in terms of the average weight for each relationship, meta-path $B - R - U - R - B$ is with higher weight than $B - R - T - R - B$, whereas in terms of total weight, meta-path $B - R - T - R - B$ is with higher weight. An interesting phenomenon is that, for Yelp-T2 task, which tries to cluster restaurants into different categories, the average weight for relationships following $B - R - U - R - B$ is 0.1716, much lower than the value (0.5864) for Yelp-T3 task, which tries to cluster shopping businesses into finer categories. This simply says that most users actually will try different kinds of food, therefore they will not be served as a good connection between restaurants as they are in other categories.

7.6 DISCUSSIONS

The Power of Meta-Path Selection Different meta-paths in heterogeneous networks could be viewed as different sources of information for defining link-based similarity between objects. There are several ways to handle different meta-paths for a mining task such as clustering: (1) to combine them at relation matrix level, such as in baselines ITC and LP; (2) to combine the clustering results at the output level, such as in ensemble baselines; and (3) to learn and improve the quality weights for each meta-path iteratively, such as in *PathSelClus*. Only the third approach is able to select different meta-paths according to different clustering tasks, whereas the other two can only output an "average" clustering result using all the information. It turns out that, in most cases, the third approach is more flexible to combine information from different sources, and its advantage has been shown in the experiment section.

Meta-Paths vs. Path Instances We now only consider the different semantics encoded by different meta-paths. In practice, different concrete paths (path instances) between two objects may also differ from each other. For example, two objects may be linked via a "bridge" or via a "hub," indicating different meanings. The difference between the two concepts: meta-path and path instance, is similar to the difference between *a source of features* and *a concrete feature* in a vector space. In this chapter, we have only discussed the selection of meta-paths. It is possible to select path instance at the object level, and the concrete method is left for future research.

[1]http://www.yelp.com/academic_dataset

CHAPTER 8

Research Frontiers

In this book, we introduced some general principles and methodologies for mining heterogeneous information networks. Although homogeneous networks are interesting subjects to study, real-world objects are usually connected via heterogeneous types of objects in complex ways, carrying critical information and rich data semantics, as shown in the examples like authors linking with papers and venues, and patients linking with diseases and treatments. Clearly, heterogeneous information networks preserve rich semantic information of the real-world data. Mining directly on heterogeneous information networks often leads to in-depth understanding of the relationships among different types of data and their regularities, models, patterns and anomalies, hence a deep insight of the networks, and fruitful mining results.

Mining heterogeneous information networks is a young and promising research field. There are many unexplored territories and challenging research issues. We illustrate a few of them here.

Constructing and Refining Heterogeneous Information Networks. Our study in most of the chapters assumes that a heterogeneous information network to be investigated contains a well-defined network schema and a large set of relatively clean and unambiguous objects and links. However, in the real world, things are more complicated.

A network extracted from a relational database may contain a well-defined schema which can be used to define the schema of its corresponding heterogeneous information network. Nevertheless, objects and links even in such a database-formed information network can still be noisy. For example, in the DBLP network, different authors may share the same name [76], that is, one node in a network may refer to multiple real-world entities; whereas in some other cases, different nodes in a network may refer to the same entity. Entity resolution will need to be integrated with network mining in order to merge and split objects or links and derive high quality results. Moreover, links in a network, roles of a node with respect to some other nodes may not be explicitly given. For example, the advisor-advisee relationship in the DBLP network [73] is not given, but such kind of relationships can be critical for understanding the growth of a research community or for some other data mining tasks. Furthermore, sometimes the connections between different nodes may not be reliable or trustable. For example, the author information for a book provided by an online book store could be erroneous or inaccurate. Multiple Web-sites may provide conflicting or compensating information for the properties of certain objects. Trustworthiness modeling [83] could be critically important for data cleaning, data integration, and quality network construction.

Construction of high-quality heterogeneous information networks becomes increasingly more challenging when we move away from relational databases towards increasingly more complicated, unstructured data, from text documents, to online web-based systems, multimedia data, and multi-lingual data. Information extraction, natural language understanding, and many other information processing techniques should be integrated with network construction and analysis techniques to ensure high-quality information networks can be constructed and progressively refined so that quality mining can be performed on better-quality heterogeneous information networks.

Notice that entity extraction, data cleaning, detection of hidden semantic relationships, and trustworthiness analysis should be integrated with the network construction and mining processes to progressively and mutually enhance the quality of construction and mining of information networks.

Diffusion analysis in heterogeneous information networks. Diffusion analysis has been studied on homogeneous networks extensively, from the innovation diffusion analysis in social science [54] to obesity diffusion in health science [12]. However, in the real world, pieces of information or diseases are propagated in more complex ways, where different types of links may play different roles. For example, diseases could propagate among people, different kinds of animals and food, via different channels. Comments on a product may propagate among people, companies, and news agencies, via traditional news feeds, social media, reviews, and so on. It is highly desirable to study the issues on information diffusion in heterogeneous information networks in order to capture the spreading models that better represent the real world patterns.

Discovery and mining of hidden information networks. Although a network can be huge, a user at a time could be only interested in a tiny portion of nodes, links, or subnetworks. Instead of directly mining the entire network, it is more fruitful to mine hidden networks "extracted" dynamically from some existing networks, based on user-specified constraints or expected node/link behaviors. For example, instead of mining an existing social network, it could be more fruitful to mine networks containing suspects and their associated links; or mine subgraphs with nontrivial nodes and high connectivity. How to discover such hidden networks and how to mine knowledge (e.g., clusters, behaviors, and anomalies) from such hidden but non-isolated networks (i.e., still intertwined with the gigantic network in both network linkages and semantics) could be an interesting but challenging problem.

Discovery of application-oriented ontological structures in heterogeneous information networks. As shown in the studies on ranking-based clustering and ranking-based classification, interconnected, multiple typed objects in a heterogeneous information network often provide critical information for generating high quality, fine-level concept hierarchies. For example, it is often difficult to identify researchers just based on their research collaboration networks. However, putting them in a heterogeneous network that links researchers with their publication, conferences, terms and research papers, their roles in the network becomes evidently clear. Moreover, people may have different preferences over ontological structures at handling different kinds of tasks. For example, some people may be interested in the research area hierarchy in the DBLP network, whereas others may be interested in finding the author lineage hierarchy.

How to incorporate user's guidance, and generate adaptable ontological structures to meet users' requirement and expectation, could be an interesting and useful topic to study.

Online analytical processing of heterogeneous information networks. The power of online analytical processing (OLAP) has been shown in multidimensional analysis of structured, relational data. Similarly, users may like to view a heterogeneous information network from different angles, in different dimension combinations, and at different levels of granularity. For example, in a bibliographic network, by specifying the object type as paper and link type as citation relation, and rolling up papers into research topics, we can immediately see the citation relationships between different research topics and figure out which research topic could be the driving force for others. However, the extension of the concept of online analysis processing (OLAP) to multi-dimensional data analysis of heterogeneous information networks is nontrivial. Not only many different applications need different ontological structures and concept hierarchies to summarize information networks but also multiple pieces of semantic information in heterogeneous networks are intertwined, determined by multiple nodes and links. There are some preliminary studies on this issue, such as [11; 72; 84], but the large territories of online analytical processing of information networks are still waiting to be explored.

Intelligent querying and semantic search in heterogeneous information networks. Given real-world data are interconnected, forming gigantic and complex heterogeneous information networks, it poses new challenges to query and search in such networks intelligently and efficiently. Given the enormous size and complexity of a large network, a user is often only interested in a small portion of the objects and links most relevant to the query. However, objects are connected and inter-dependent on each other, how to search effectively in a large network for a given user's query could be a challenge. Similarity search that returns the most similar objects to a queried object, as studied in this book [65] and its follow-up [57], will serve as a basic function for semantic search in heterogeneous networks. Such a kind of similarity search may lead to useful applications, such as product search in e-commerce networks and patent search in patent networks.

Search functions should be further enhanced and integrated with many other functions. For example, structural search [78], which tries to find semantically similar structures given a structural query, may be useful for finding pattern in an e-commerce network involving buyers, sellers, products, and their interactions. Also, a recommendation system may take advantage of heterogeneous information networks that link among products, customers and their properties to make improved recommendations. Querying and semantic search in heterogeneous information networks opens another interesting frontier on research related to mining heterogeneous information networks.

Bibliography

[1] C. C. Aggarwal (ed.), *Social Network Data Analytics*. Springer, 2011. DOI: 10.1007/978-1-4419-8462-3 Cited on page(s) 4

[2] E. M. Airoldi, D. M. Blei, S. E. Fienberg, and E. P. Xing. Mixed membership stochastic blockmodels. *J. Mach. Learn. Res.*, 9:1981–2014, 2008. DOI: 10.1145/1390681.1442798 Cited on page(s) 98

[3] A. Banerjee, S. Merugu, I. S. Dhillon, and J. Ghosh. Clustering with bregman divergences. *J. Mach. Learn. Res.*, 6:1705–1749, 2005. Cited on page(s) 129

[4] A.-L. Barabási and R. Albert. Emergence of scaling in random networks. *Science*, 286(5439):509–512, 1999. DOI: 10.1126/science.286.5439.509 Cited on page(s) 19, 29

[5] A.-L. Barabási, H. Jeong, Z. Neda, E. Ravasz, A. Schubert, and T. Vicsek. Evolution of the social network of scientific collaborations. *arXiv:cond-mat/0104162v1*, 2001. DOI: 10.1016/S0378-4371(02)00736-7 Cited on page(s) 74

[6] S. Basu, A. Banerjee, and R. Mooney. Semi-supervised clustering by seeding. In *Proc. 2002 Int. Conf. Machine Learning (ICML'02)*, Sydney, Australia, July 2002. Cited on page(s) 129

[7] M. Belkin, P. Niyogi, and V. Sindhwani. Manifold regularization: A geometric framework for learning from examples. *J. Mach. Learn. Res.*, 7:2399–2434, 2006. Cited on page(s) 38

[8] J. Bilmes. A gentle tutorial on the EM algorithm and its application to parameter estimation for Gaussian Mixture and Hidden Markov

Models. *University of Berkeley Tech Rep ICSITR*, (ICSI-TR-97-021), 1997. Cited on page(s) 19, 107

[9] A. P. Bradley. The use of the area under the ROC curve in the evaluation of machine learning algorithms. *Pattern Recognition*, 30(7):1145–1159, 1997. DOI: 10.1016/S0031-3203(96)00142-2 Cited on page(s) 79

[10] S. Brin and L. Page. The anatomy of a large-scale hypertextual web search engine. *Computer Networks*, 30(1-7):107–117, 1998. DOI: 10.1016/S0169-7552(98)00110-X Cited on page(s) 1, 11, 14, 44

[11] C. Chen, X. Yan, F. Zhu, J. Han, and P. S. Yu. Graph OLAP: Towards online analytical processing on graphs. In *Proc. 2008 Int. Conf. Data Mining (ICDM'08)*, Pisa, Italy, December 2008. DOI: 10.1109/ICDM.2008.30 Cited on page(s) 137

[12] N. A. Christakis and J. H. Fowler. The spread of obesity in a large social network over 32 years. *The New England Journal of Medicine*, 357(4):370–379, 2007. DOI: 10.1056/NEJMsa066082 Cited on page(s) 136

[13] F. R. K. Chung. *Spectral Graph Theory. Regional Conference Series in Mathematics*. No. 92, American Mathematical Society, 1997. Cited on page(s) 43

[14] A. Clauset, M. E. J. Newman, and C. Moore. Finding community structure in very large networks. In *Phys. Rev. E*, Vol.70, No.6, 066111, 2004. DOI: 10.1103/PhysRevE.70.066111 Cited on page(s) 98

[15] D. Cohn and H. Chang. Learning to probabilistically identify authoritative documents. In *Proc. 2000 Int. Conf. Machine Learning (ICML'00)*, Stanford, CA, June 2000. Cited on page(s) 122

[16] D. R. Cutting, D. R. Karger, J. O. Pedersen, and J. W. Tukey. Scatter/gather: a cluster-based approach to browsing large document collections. In *Proc. 1992 ACM Int. Conf. Research and Development in Information Retrieval (SIGIR '92)*, Copenhagen, Denmark, June 1992. DOI: 10.1145/133160.133214 Cited on page(s) 14, 45

[17] A. P. Dempster, N. M. Laird, and D. B. Rubin. Maximum likelihood from incomplete data via the em algorithm. *Journal of The Royal Statistical Society*, SERIES B, 39(1):1–38, 1977. Cited on page(s) 107

[18] I. S. Dhillon, S. Mallela, and R. Kumar. A divisive information-theoretic feature clustering algorithm for text classification. *J. Mach. Learn. Res.*, 3:1265–1287, 2003. Cited on page(s) 129

[19] A. J. Dobson. *An Introduction to Generalized Linear Models*, 2nd Edition. Chapman & Hall/CRC, 2001. DOI: 10.1201/9781420057683 Cited on page(s) 83, 86

[20] M. Faloutsos, P. Faloutsos, and C. Faloutsos. On power-law relationships of the internet topology. In *Proc. 1999 ACM Conf. Applications, Technologies, Architectures, and Protocols for Computer Communication (SIGCOMM'99)*, Cambridge, MA, August 1999. DOI: 10.1145/316194.316229 Cited on page(s) 29

[21] J. Gao, W. Fan, Y. Sun, and J. Han. Heterogeneous source consensus learning via decision propagation and negotiation. In *Proc. 2009 ACM Int. Conf. Knowledge Discovery and Data Mining (KDD'09)*, Paris, France, June 2009. DOI: 10.1145/1557019.1557061 Cited on page(s) 110

[22] C. L. Giles. The future of citeseer: *citeseer^x*. In *Proc. 10th European Conf. Principles and Practice of Knowledge Discovery in Databases (PKDD'06)*, Berlin, Germany, September 2006. DOI: 10.1007/11871637_2 Cited on page(s) 1

[23] Z. Gyöngyi, H. Garcia-Molina, and J. Pedersen. Combating web spam with trustrank. In *Proc. 2004 Int. Conf. Very Large Data Bases (VLDB'04)*, Toronto, Canada, August 2004. Cited on page(s) 17

[24] M. A. Hasan, V. Chaoji, S. Salem, and M. Zaki. Link prediction using supervised learning. In *Proc. of SDM'06 workshop on Link Analysis, Counterterrorism and Security*, Bethesda, MD, April 2006. Cited on page(s) 73

[25] T. Hofmann. Probabilistic latent semantic analysis. In *Proc. 15th Annual Conf. Uncertainty in Artificial Intelligence (UAI'99)*, Stockholm, Sweden, July 1999. DOI: 10.1145/312624.312649 Cited on page(s) 35, 103

[26] T. Hofmann. Probabilistic latent semantic indexing. In *Proc. 1999 ACM Int. Conf. Research and Development in Information Retrieval (SIGIR'99)*, Berkeley, CA, August 1999. DOI: 10.1145/312624.312649 Cited on page(s) 122

[27] K. Jarvelin and J. Kekalainen. Cumulated gain-based evaluation of ir techniques. *ACM Transactions on Information Systems*, 20(4), 2003. DOI: 10.1145/582415.582418 Cited on page(s) 69

[28] G. Jeh and J. Widom. Simrank: a measure of structural-context similarity. In *Proc. 2002 ACM Int. Conf. Knowledge Discovery in Databases (KDD'02)*, Edmonton,

Canada, July 2002. DOI: 10.1145/775047.775126 Cited on page(s) 13, 24, 57

[29] G. Jeh and J. Widom. Scaling personalized web search. In *Proc. 2003 Int. World Wide Web Conf. (WWW'03)*, Budapest, Hungary, May 2003. DOI: 10.1145/775152.775191 Cited on page(s) 57

[30] M. Ji, J. Han, and M. Danilevsky. Ranking-based classification of heterogeneous information networks. In *Proc. 2011 ACM Int. Conf. on Knowledge Discovery and Data Mining (KDD'11)*, San Diego, CA, August 2011. DOI: 10.1145/2020408.2020603 Cited on page(s) 5, 6

[31] M. Ji, Y. Sun, M. Danilevsky, J. Han, and J. Gao. Graph regularized transductive classification on heterogeneous information networks. In *Proc. 2010 European Conf. Machine Learning and Principles and Practice of Knowledge Discovery in Databases (ECML PKDD'10)*, Barcelona, Spain, September 2010. DOI: 10.1007/978-3-642-15880-3_42 Cited on page(s) 5, 6, 47, 48

[32] W. Jiang, J. Vaidya, Z. Balaporia, C. Clifton, and B. Banich. Knowledge discovery from transportation network data. In *Proc. 2005 Int. Conf. Data Engineering (ICDE'05)*, Tokyo, Japan, April 2005. DOI: 10.1109/ICDE.2005.82 Cited on page(s) 1

[33] L. Katz. A new status index derived from sociometric analysis. *Psychometrika*, 18(1):39–43, 1953. DOI: 10.1007/BF02289026 Cited on page(s) 74, 75

[34] J. M. Kleinberg. Authoritative sources in a hyperlinked environment. *J. ACM*, 46(5):604–632, 1999. DOI: 10.1145/324133.324140 Cited on page(s) 11, 14, 44

[35] M. Kolahdouzan and C. Shahabi. Voronoi-based k nearest neighbor search for spatial network databases. In *Proc. 2004 Int. Conf. Very Large Data Bases (VLDB'04)*, Toronto, Canada, August 2004. Cited on page(s) 57

[36] B. Kulis, S. Basu, I. Dhillon, and R. Mooney. Semi-supervised graph clustering: a kernel approach. In *Proc. 22nd Int. Conf. Machine Learning (ICML'05)*, Bonn, Germany, August 2005. DOI: 10.1145/1102351.1102409 Cited on page(s) 118

[37] D. D. Lee and H. S. Seung. Algorithms for non-negative matrix factorization. In *Proc. 2000 Neural Info. Processing Systems Conf. (NIPS'00)*, Denver, CO, 2000. Cited on page(s) 126

[38] V. Leroy, B. B. Cambazoglu, and F. Bonchi. Cold start link prediction. In *Proc. 2010 ACM Int. Conf. Knowledge Discovery and Data Mining (KDD'10)*, Washington D.C., July 2010. DOI: 10.1145/1835804.1835855 Cited on page(s) 73

[39] D. Liben-Nowell and J. Kleinberg. The link prediction problem for social networks. In *Proc. 2003 Int. Conf. Information and Knowledge Management (CIKM'03)*, New Orleans, LA, November 2003. DOI: 10.1145/956863.956972 Cited on page(s) 73, 74, 79

[40] R. N. Lichtenwalter, J. T. Lussier, and N. V. Chawla. New perspectives and methods in link prediction. In *Proc. 2010 ACM Int. Conf. Knowledge Discovery and Data Mining (KDD'10)*, Washington D.C., July 2010. DOI: 10.1145/1835804.1835837 Cited on page(s) 73, 75, 76

[41] B. Long, Z. M. Zhang, X. Wu, and P. S. Yu. Spectral clustering for multi-type relational data. In *Proc. 23nd Int. Conf. Machine Learning (ICML'06)*, Pittsburgh, Pennsylvania, June 2006. DOI: 10.1145/1143844.1143918 Cited on page(s) 38, 42, 48

[42] B. Long, Z. M. Zhang, and P. S. Yu. A probabilistic framework for relational clustering. In *Proc. 2007 Int. Conf. Knowledge Discovery and Data Mining (KDD'07)*, San Jose, CA, August 2007. DOI: 10.1145/1281192.1281244 Cited on page(s) 98

[43] Q. Lu and L. Getoor. Link-based classification. In *Proc. 2003 Int. Conf. Machine Learning (ICML'03)*, Washington D.C., August 2003. DOI: 10.1145/956863.956938 Cited on page(s) 38

[44] U. Luxburg. A tutorial on spectral clustering. *Statistics and Computing*, 17:395–416, 2007. DOI: 10.1007/s11222-007-9033-z Cited on page(s) 13, 98

[45] S. A. Macskassy and F. Provost. A simple relational classifier. In *Proceedings of the Second Workshop on Multi-Relational Data Mining (MRDM-2003) at KDD-2003*, Washington D.C., August 2003. Cited on page(s) 38, 41, 51

[46] S. A. Macskassy and F. Provost. Classification in networked data: A toolkit and a univariate case study. *J. Mach. Learn. Res.*, 8:935–983, 2007. Cited on page(s) 38, 44, 51

[47] Q. Mei, D. Cai, D. Zhang, and C. Zhai. Topic modeling with network regularization. In *Proc. 17th Int. World Wide Web Conf. (WWW'08)*, Beijing, China, April 2008. DOI: 10.1145/1367497.1367512 Cited on page(s) 98, 111

[48] J. Neville and D. Jensen. Relational dependency networks. *J. Mach. Learn. Res.*, 8:653–692, 2007. Cited on page(s) 38

[49] M. E. J. Newman. Clustering and preferential attachment in growing networks. *Physical Review Letters E*, 64, 2001. DOI: 10.1103/PhysRevE.64.025102 Cited

on page(s) 74

[50] M. E. J. Newman. Assortative mixing in networks. *Physical Review Letters*, 89(20):208701+, 2002. DOI: 10.1103/PhysRevLett.89.208701 Cited on page(s) 29

[51] Z. Nie, Y. Zhang, J.-R. Wen, and W.-Y. Ma. Object-level ranking: bringing order to web objects. In *Proc. 2005 Int. World Wide Web Conf. (WWW'05)*, Chiba, Japan, May 2005. DOI: 10.1145/1060745.1060828 Cited on page(s) 16, 28

[52] A. Popescul, R. Popescul, and L. H. Ungar. Statistical relational learning for link prediction. In *Proc. of the Workshop on Learning Statistical Models from Relational Data at IJCAI-2003*, Acapulco, Mexico, August 2003. Cited on page(s) 90

[53] K. Punera and J. Ghosh. Consensus-based ensembles of soft clusterings. *Appl. Artif. Intell.*, 22:780–810, 2008. DOI: 10.1080/08839510802170546 Cited on page(s) 129

[54] E. M. Rogers. *Diffusion of Innovations*. Free Press, 5th edition, 2003. Cited on page(s) 136

[55] S. Roy, T. Lane, and M. Werner-Washburne. Integrative construction and analysis of condition-specific biological networks. In *Proc. 2007 Conf. on Artificial Intelligence (AAAI'07)*, Vancouver, BC, July 2007. Cited on page(s) 1

[56] P. Sen and L. Getoor. Link-based classification. Technical Report CS-TR-4858, University of Maryland, February 2007. Cited on page(s) 38, 44, 51

[57] C. Shi, X. Kong, P. S. Yu, S. Xie, and B. Wu. Relevance search in heterogeneous networks. In *Proc. 2012 Int. Conf. on Extending Data Base Technology (EDBT'12)*, Berlin, Germany, March 2012. DOI: 10.1145/2247596.2247618 Cited on page(s) 137

[58] J. Shi and J. Malik. Normalized cuts and image segmentation. *IEEE Transactions on Pattern Analysis and Machine Intelligence*, 22:888–905, 1997. DOI: 10.1109/34.868688 Cited on page(s) 13, 24

[59] M. Shiga, I. Takigawa, and H. Mamitsuka. A spectral clustering approach to optimally combining numericalvectors with a modular network. In *Proc. 2007 Int. Conf. Knowledge Discovery and Data Mining (KDD'07)*, San Jose, CA, August 2007. DOI: 10.1145/1281192.1281262 Cited on page(s) 98, 113

[60] A. Strehl and J. Ghosh. Cluster ensembles—a knowledge reuse framework for combining multiple partitions. *J. Mach. Learn. Res.*, 3:583–617, 2003. DOI: 10.1162/153244303321897735 Cited on page(s) 24, 111

[61] Y. Sun, C. C. Aggarwal, and J. Han. Relation strength-aware clustering of heterogeneous information networks with incomplete attributes. In *Proc. 2012 Int. Conf. Very Large Data Bases (VLDB'12)*, Istanbul, Turkey, August 2012. Cited on page(s) 5, 7, 107

[62] Y. Sun, R. Barber, M. Gupta, C. C. Aggarwal, and J. Han. Co-author relationship prediction in heterogeneous bibliographic networks. In *Proc. 2011 Int. Conf. Advances in Social Network Analysis and Mining (ASONAM'11)*, Kaohsiung, Taiwan, July 2011. DOI: 10.1109/ASONAM.2011.112 Cited on page(s) 5, 7

[63] Y. Sun, J. Han, C.C. Aggarwal, and N.V. Chawla. When will it happen?-relationship prediction in heterogeneous information networks. In *Proc. 2012 ACM Int. Conf. on Web Search and Data Mining (WSDM'12)*, Seattle, WA, Feb. 2012. DOI: 10.1145/2124295.2124373 Cited on page(s) 5, 7

[64] Y. Sun, J. Han, J. Gao, and Y. Yu. Itopicmodel: Information network-integrated topic modeling. In *Proc. 2009 Int. Conf. Data Mining (ICDM'09)*, Miami, FL, December 2009. DOI: 10.1109/ICDM.2009.43 Cited on page(s) 68, 111

[65] Y. Sun, J. Han, X. Yan, P. S. Yu, and T. Wu. PathSim: Meta path-based top-k similarity search in heterogeneous information networks. In *Proc. 2011 Int. Conf. on Very Large Data Bases (VLDB'11)*, Seattle, WA, August 2011. Cited on page(s) 5, 6, 66, 76, 83, 137

[66] Y. Sun, J. Han, P. Zhao, Z. Yin, H. Cheng, and T. Wu. RankClus: integrating clustering with ranking for heterogeneous information network analysis. In *Proc. 2009 Int. Conf. Extending Data Base Technology (EDBT'09)*, Saint-Petersburg, Russia, March 2009. DOI: 10.1145/1516360.1516426 Cited on page(s) 5, 6, 16, 38, 63, 122

[67] Y. Sun, B. Norick, J. Han, X. Yan, P. S. Yu, and X. Yu. Integrating meta-path selection with user-guided object clustering in heterogeneous information networks. In *Proc. 2012 ACM Int. Conf. on Knowledge Discovery and Data Mining (KDD'12)*, Beijing, China, August 2012. Cited on page(s) 5, 7

[68] Y. Sun, J. Tang, J. Han, M. Gupta, and B. Zhao. Community evolution detection in dynamic heterogeneous information networks. In *Proceedings of the Eighth Workshop on Mining and Learning with Graphs MLG'10 at KDD'10,*

Washington D.C., August 2010. DOI: 10.1145/1830252.1830270 Cited on page(s) 5

[69] Y. Sun, Y. Yu, and J. Han. Ranking-based clustering of heterogeneous information networks with star network schema. In *Proc. 2009 ACM Int. Conf. Knowledge Discovery and Data Mining (KDD'09)*, Paris, France, June 2009. DOI: 10.1145/1557019.1557107 Cited on page(s) 5, 6, 39, 98, 110, 128

[70] B. Taskar, P. Abbeel, and D. Koller. Discriminative probabilistic models for relational data. In *Proc. 2002 Annual Conf. Uncertainty in Artificial Intelligence (UAI'02)*, Edmonton, Canada, 2002. Cited on page(s) 38

[71] B. Taskar, E. Segal, and D. Koller. Probabilistic classification and clustering in relational data. In *Proc. 17th Intl. Joint Conf. Artificial Intelligence (IJCAI'01)*, Seattle, WA, 2001. Cited on page(s) 98

[72] Y. Tian, R. A. Hankins, and J. M. Patel. Efficient aggregation for graph summarization. In *Proc. 2008 ACM Int. Conf. Management of Data (SIGMOD'08)*, Vancouver, Canada, June 2008. DOI: 10.1145/1376616.1376675 Cited on page(s) 137

[73] C. Wang, J. Han, Y. Jia, J. Tang, D. Zhang, Y. Yu, and J. Guo. Mining advisor-advisee relationships from research publication networks. In *Proc. 2010 ACM Int. Conf. Knowledge Discovery and Data Mining (KDD'10)*, Washington D.C., July 2010. DOI: 10.1145/1835804.1835833 Cited on page(s) 135

[74] C. Wang, V. Satuluri, and S. Parthasarathy. Local probabilistic models for link prediction. In *Proc. 2007 Int. Conf. Data Mining (ICDM'07)*, Omaha, NE, October 2007. DOI: 10.1109/ICDM.2007.108 Cited on page(s) 73

[75] T. Yang, R. Jin, Y. Chi, and S. Zhu. Combining link and content for community detection: a discriminative approach. In *Proc. 2009 ACM Int. Conf. Knowledge Discovery and Data Mining (KDD'09)*, Paris, France, June 2009. DOI: 10.1145/1557019.1557120 Cited on page(s) 98

[76] X. Yin, J. Han, and P. Yu. Object distinction: Distinguishing objects with identical names. In *Proc. 2007 Int. Conf. Data Engineering (ICDE'07)*, Istanbul, Turkey, April 2007. DOI: 10.1109/ICDE.2007.368983 Cited on page(s) 135

[77] Z. Yin, R. Li, Q. Mei, and J. Han. Exploring social tagging graph for web object classification. In *Proc. 2009 ACM Int. Conf. Knowledge Discovery and Data Mining (KDD'09)*, Paris, France, June 2009. DOI: 10.1145/1557019.1557123 Cited on page(s) 38

[78] X. Yu, Y. Sun, P. Zhao, and J. Han. Query-driven discovery of semantically similar substructures in heterogeneous networks (system demo). In *Proc. 2012 ACM Int. Conf. on Knowledge Discovery and Data Mining (KDD'12)*, Beijing, China, August 2012. Cited on page(s) 137

[79] O. Zamir and O. Etzioni. Grouper: A dynamic clustering interface to web search results. *Computer Networks*, 31(11-16):1361–1374, 1999. DOI: 10.1016/S1389-1286(99)00054-7 Cited on page(s) 14, 45

[80] C. H. Zha, H. Zha, X. He, C. Ding, H. Simon, and M. Gu. Spectral relaxation for k-means. In *Proc. 2001 Neural Info. Processing Systems Conf. (NIPS'01)*, Vancouver, Canada, December 2001. Cited on page(s) 113

[81] C. Zhai. *Statistical Language Models for Information Retrieval*. Now Publishers Inc., Hanover, MA, USA, 2008. Cited on page(s) 111

[82] C. Zhai and J. D. Lafferty. A study of smoothing methods for language models applied to information retrieval. *ACM Transactions on Information Systems*, 22(2):179–214, 2004. DOI: 10.1145/984321.984322 Cited on page(s) 31

[83] B. Zhao, B. I. P. Rubinstein, J. Gemmell, and J. Han. A Bayesian approach to discovering truth from conflicting sources for data integration. In *Proc. 2012 Int. Conf. Very Large Data Bases (VLDB'12)*, Istanbul, Turkey, August 2012. Cited on page(s) 135

[84] P. Zhao, X. Li, D. Xin, and J. Han. Graph cube: On warehousing and OLAP multidimensional networks. In *Proc. of 2011 ACM Int. Conf. on Management of Data (SIGMOD'11)*, Athens, Greece, June 2011. DOI: 10.1145/1989323.1989413 Cited on page(s) 137

[85] D. Zhou, O. Bousquet, T. N. Lal, J. Weston, and B. Schölkopf. Learning with local and global consistency. In *Advances in Neural Information Processing Systems 16*, pages 321–328, 2004, MIT Press. Cited on page(s) 38, 41, 42, 43, 47, 51

[86] D. Zhou, J. Weston, A. Gretton, O. Bousquet, and B. Schölkopf. Ranking on data manifolds. In *Advances in Neural Information Processing Systems 16*, 2004, MIT Press. Cited on page(s) 17, 29, 48

[87] Y. Zhou, H. Cheng, and J. X. Yu. Graph clustering based on structural/attribute similarities. *Proc. VLDB Endowment*, 2(1):718–729, 2009. Cited on page(s) 98

[88] X. Zhu and Z. Ghahramani. Learning from labeled and unlabeled data with label propagation. Technical Report CMU-CALD-02-107, Carnegie Mellon University,

2002. Cited on page(s) 118

[89] X. Zhu, Z. Ghahramani, and J. D. Lafferty. Semi-Supervised learning using gaussian fields and harmonic functions. In *Proc. 2003 Int. Conf. Machine Learning (ICML'03)*, Washington D.C., August 2003. Cited on page(s) 38, 118, 129

Authors' Biographies

YIZHOU SUN

Yizhou Sun received her Ph.D. in Computer Science from the University of Illinois at Urbana-Champaign in 2012. She will be an assistant professor in the College of Computer and Information Science at Northeastern University. Her principal research interest is in mining information and social networks, and more generally in data mining, database systems, statistics, machine learning, information retrieval, and network science, with a focus on modeling novel problems and proposing scalable algorithms for large-scale, real-world applications. Yizhou has over 30 publications in books, journals, and major conferences. Tutorials based on her thesis work on mining heterogeneous information networks have been given in several premier conferences, including EDBT 2009, SIGMOD 2010, KDD 2010, ICDE 2012, VLDB 2012, and ASONAM 2012. She received ACM KDD 2012 Best Student Paper Award.

JIAWEI HAN

Jiawei Han is the Abel Bliss Professor of Computer Science at the University of Illinois at Urbana-Champaign. His research includes data mining, information network analysis, database systems, and data warehousing, with over 600 journal and conference publications. He has chaired or served on many program committees of international conferences, including PC co-chair for KDD, SDM, and ICDM conferences, and Americas Coordinator for VLDB conferences. He also served as the founding Editor-In-Chief of ACM Transactions on Knowledge Discovery from Data and is serving as the Director of Information Network Academic Research Center supported by U.S. Army Research Lab. He is Fellow of ACM and Fellow of IEEE, and received 2004 ACM SIGKDD Innovations Award, 2005 IEEE Computer Society Technical Achievement Award, 2009 IEEE Computer Society Wallace McDowell Award, and 2011 Daniel C. Drucker Eminent Faculty Award at UIUC. His book, *Data Mining: Concepts and Techniques*, has been used popularly as a textbook worldwide.

Printed in the United States
by Baker & Taylor Publisher Services